Giuliano Benenti. Born in Voghera (Pavia), Italy, November 7, 1969. He is a researcher in Theoretical Physics at Università dell' Insubria, Como. He received his Ph.D. in physics at Universita di Milano, Italy and was a postdoctoral fellow at CEA, Saclay, France. His main research interests are in the fields of classical and quantum chaos, open quantum systems, mesoscopic physics, disordered systems, phase transitions, many-body systems and quantum information theory.

Giulio Casati. Born in Brenna (Como), Italy, December 9, 1942. He is a professor of Theoretical Physics at Università dell' Insubria, Como, former professor at Milano University, and distinguished visiting professor at NUS, Singapore. A member of the Academia Europea, and director of the Center for Nonlinear and Complex Systems, he was awarded the F. Somaini Italian prize for physics in 1991. As editor of several volumes on classical and quantum chaos, he has done pioneering research in nonlinear dynamics, classical and quantum chaos with applications to atomic, solid state, nuclear physics and, more recently, to quantum computers.

Giuliano Strini. Born in Roma, Italy, September 9, 1937. He is an associate professor in Experimental Physics and has been teaching a course on Quantum Computation at Universita di Milano, for several years. From 1963, he has been involved in the construction and development of the Milan Cyclotron. His publications concern nuclear reactions and spectroscopy, detection of gravitational waves, quantum optics and, more recently, quantum computers. He is a member of the Italian Physical Society, and also the Optical Society of America.

Principles of Quantum Computation and Information

Volume I: Basic Concepts

Giuliano Benenti and Giulio Casati

Università degli Studi dell Insubria, Italy
Istituto Nazionale per la Fisica della Materia, Italy

Giuliano Strini

Università di Milano, Italy

World Scientific

NEW JERSEY · LONDON · SINGAPORE · BEIJING · SHANGHAI · HONG KONG · TAIPEI · CHENNAI

Published by

World Scientific Publishing Co. Pte. Ltd.

5 Toh Tuck Link, Singapore 596224

USA office: 27 Warren Street, Suite 401-402, Hackensack, NJ 07601

UK office: 57 Shelton Street, Covent Garden, London WC2H 9HE

British Library Cataloguing-in-Publication Data
A catalogue record for this book is available from the British Library.

First published 2004
Reprinted 2005, 2008

PRINCIPLES OF QUANTUM COMPUTATION AND INFORMATION
Volume I: Basic Concepts

ISBN-13 978-981-238-830-8
ISBN-10 981-238-830-3
ISBN-13 978-981-238-858-2 (pbk)
ISBN-10 981-238-858-3 (pbk)

Printed in Singapore by B & JO Enterprise

To Silvia
g.b.

To my wife for her love and encouragement
g.c.

To my family and friends
g.s.

Preface

Purpose of the book

This book is addressed to undergraduate and graduate students in physics, mathematics and computer science. It is written at a level comprehensible to readers with the background of a student near to the end of an undergraduate course in one of the above three disciplines. Note that no prior knowledge either of quantum mechanics or of classical computation is required to follow this book. Indeed, the first two chapters are a simple introduction to classical computation and quantum mechanics. Our aim is that these chapters should provide the necessary background for an understanding of the subsequent chapters.

The book is divided into two volumes. In volume I, after providing the necessary background material in classical computation and quantum mechanics, we develop the basic principles and discuss the main results of quantum computation and information. Volume I would thus be suitable for a one-semester introductory course in quantum information and computation, for both undergraduate and graduate students. It is also our intention that volume I be useful as a general education for other readers who would like to learn the basic principles of quantum computation and information and who have the basic background in physics and mathematics acquired in undergraduate courses in physics, mathematics or computer science.

Volume II deals with various important aspects, both theoretical and experimental, of quantum computation and information. This volume necessarily contains parts that are more technical or specialized. For its understanding, a knowledge of the material discussed in the first volume is necessary.

General approach

Quantum computation and information is a new and rapidly developing field. It is therefore not easy to grasp the fundamental concepts and central results without having to face many technical details. Our purpose in this book is to provide the reader interested in this field with a useful and not overly heavy guide. Therefore, mathematical rigour is not our primary concern. Instead, we have tried to present a simple and systematic treatment, such that the reader might understand the material presented without the need for consulting other texts. Moreover, we have not tried to cover all aspects of the field, preferring to concentrate on the fundamental concepts. Nevertheless, the two volumes should prove useful as a reference guide to researchers just starting out in the field.

To fully familiarize oneself with the subject, it is important to practice solving problems. The book contains a large number of exercises (with solutions), which are an essential complement to the main text. In order to develop a solid understanding of the arguments dealt with here, it is indispensable that the student try to solve a large part of them.

Note to the reader

Some of the material presented is not necessary for understanding the rest of the book and may be omitted on a first reading. We have adopted two methods of highlighting such parts:

1) The sections or subsections with an asterisk before the title contain more advanced or complementary material. Such parts may be omitted without risk of encountering problems in reading the rest of the book.

2) Comments, notes or examples are printed in a small typeface.

Acknowledgments

We are indebted to several colleagues for criticism and suggestions. In particular, we wish to thank Alberto Bertoni, Gabriel Carlo, Rosario Fazio, Bertrand Georgeot, Luigi Lugiato, Sandro Morasca, Simone Montangero, Massimo Palma, Saverio Pascazio, Nicoletta Sabadini, Marcos Saraceno, Stefano Serra Capizzano and Robert Walters, who read preliminary versions of the book. We are also grateful to Federico Canobbio and Sisi Chen. Special thanks is due to Philip Ratcliffe, for useful remarks and suggestions, which substantially improved our book. Obviously no responsibility should be attributed to any of the above regarding possible flaws that might remain, for which the authors alone are to blame.

Contents

Contents of Volume II

About the Cover

This acrostic is the famous *sator* formula. It can be translated as:

'*Arepo the sower holds the wheels at work*'

The text may be read in four different ways:

(i)   horizontally, from left to right (downward) and from right to left (upward);
(ii) vertically, downward (left to right) and upward (right to left).

The resulting phrase is always the same.

It has been suggested that it might be a form of secret message.

This acrostic was unearthed during archeological excavation work at Pompeii, which was buried, as well known, by the eruption of Vesuvius in 79 A.D. The formula can be found throughout the Roman Empire, probably also spread by legionnaires. Moreover, it has been found in Mesopotamia, Egypt, Cappadocia, Britain and Hungary.

The *sator* acrostic may have a mystical significance and might have been used as a means for persecuted Christians to recognize each other (it can be rearranged into the form of a cross, with the opening words of the Lord's prayer, *A Paternoster O*, both vertically and horizontally, intersecting at the letter N, the Latin letters A and O corresponding to the Greek letters alpha and omega, beginning and end of all things).

Introduction

Quantum mechanics has had an enormous technological and societal impact. To appreciate this point, it is sufficient to consider the invention of the transistor, perhaps the most remarkable among the countless other applications of quantum mechanics. On the other hand, it is also easy to see the enormous impact of computers on everyday life. The importance of computers is such that it is appropriate to say that we are now living in the *information age*. This information revolution became possible thanks to the invention of the transistor, that is, thanks to the synergy between computer science and quantum physics.

Today this synergy offers completely new opportunities and promises exciting advances in both fundamental science and technological application. We are referring here to the fact that **quantum mechanics can be used to process and transmit information**.

Miniaturization provides us with an intuitive way of understanding why, in the near future, quantum laws will become important for computation. The electronics industry for computers grows hand-in-hand with the decrease in size of integrated circuits. This miniaturization is necessary to increase computational power, that is, the number of floating-point operations per second (flops) a computer can perform. In the 1950's, electronic computers based on vacuum-tube technology were capable of performing approximately 10^3 floating-point operations per second, while nowadays there exist supercomputers whose power is greater than 10 teraflops (10^{13} flops). As we have already remarked, this enormous growth of computational power has been made possible owing to progress in miniaturization, which may be quantified empirically in Moore's law. This law is the result of a remarkable observation made by Gordon Moore in 1965: the number

1

of transistors that may be placed on a single integrated-circuit chip doubles approximately every $18 - 24$ months. This exponential growth has not yet saturated and Moore's law is still valid. At the present time the limit is approximately 10^8 transistors per chip and the typical size of circuit components is of the order of 100 nanometres. Extrapolating Moore's law, one would estimate that around the year 2020 we shall reach the atomic size for storing a single bit of information. At that point, quantum effects will become unavoidably dominant.

It is clear that, besides quantum effects, other factors could bring Moore's law to an end. In the first place, there are economic considerations. Indeed, the cost of building fabrication facilities to manufacture chips has also increased exponentially with time. Nevertheless, it is important to understand the ultimate limitations set by quantum mechanics. Even though we might overcome economic barriers by means of technological breakthroughs, quantum physics sets fundamental limitations on the size of the circuit components. The first question under debate is whether it would be more convenient to push the silicon-based transistor to its physical limits or instead to develop alternative devices, such as quantum dots, single-electron transistors or molecular switches. A common feature of all these devices is that they are on the nanometre length scale and therefore quantum effects play a crucial role.

So far, we have talked about quantum switches that could substitute silicon-based transistors and possibly be connected together to execute classical algorithms based on Boolean logic. In this perspective, quantum effects are simply unavoidable corrections that must be taken into account owing to the nanometre size of the switches. A quantum computer represents a radically different challenge: the aim is to build a machine *based on quantum logic*, that is, it processes the information and performs logic operations by exploiting the laws of quantum mechanics.

The unit of quantum information is known as a *qubit* (the quantum counterpart of the classical *bit*) and a quantum computer may be viewed as a many-qubit system. Physically, a qubit is a two-level system, like the two spin states of a spin-$\frac{1}{2}$ particle, the vertical and horizontal polarization states of a single photon or the ground and excited states of an atom. A quantum computer is a system of many qubits, whose evolution can be controlled, and a quantum computation is a unitary transformation that acts on the many-qubit state describing the quantum computer.

The power of quantum computers is due to typical quantum phenomena, such as the *superposition* of quantum states and *entanglement*. There is an inherent quantum parallelism associated with the superposition principle. In simple terms, a quantum computer can process a large number of classical inputs in a single run. On the other hand, this implies a large number of possible outputs. It is the task of quantum algorithms, which are based on quantum logic, to exploit the inherent quantum parallelism of quantum mechanics to highlight the desired output. In short, to be useful, quantum computers require the development of appropriate quantum software, that is, of efficient quantum algorithms.

In the 1980's Feynman suggested that a quantum computer based on quantum logic would be ideal for simulating quantum-mechanical systems and his ideas have spawned an active area of research in physics. It is also remarkable that quantum mechanics can help in the solution of basic problems of computer science. In 1994, Peter Shor proposed a quantum algorithm that efficiently solves the prime-factorization problem: given a composite integer, find its prime factors. This is a central problem in computer science and it is conjectured, though not proven, that for a classical computer it is computationally difficult to find the prime factors. Shor's algorithm efficiently solves the integer factorization problem and therefore it provides an exponential improvement in speed with respect to any known classical algorithm. It is worth mentioning here that there are cryptographic systems, such as RSA, that are used extensively today and that are based on the conjecture that no efficient algorithms exist for solving the prime factorization problem. Hence, Shor's algorithm, if implemented on a large-scale quantum computer, would break the RSA cryptosystem. Lov Grover has shown that quantum mechanics can also be useful for solving the problem of searching for a marked item in an unstructured database. In this case, the gain with respect to classical computation is quadratic.

Another interesting aspect of the quantum computer is that, in principle, it avoids dissipation. Present day classical computers, which are based on irreversible logic operations (gates), are *intrinsically* dissipative. The minimum energy requirements for irreversible computation are set by Landauer's principle: each time a single bit of information is erased, the amount of energy dissipated into the environment is at least $k_B T \ln 2$, where k_B is Boltzmann's constant and T the temperature of the environment surrounding the computer. Each irreversible classical gate must dissipate at least this amount of energy (in practice, present-day computers dissipate more by orders of magnitude). In contrast, quantum evolution is unitary and thus quantum logic gates must be reversible. Therefore, at least in principle, there is no energy dissipation during a quantum computer run.

It is well known that a small set of elementary logic gates allows the implementation of any complex computation on a classical computer. This is very important: it means that, when we change the problem, we do not need to modify our computer hardware. Fortunately, the same property remains valid for a quantum computer. It turns out that, in the quantum circuit model, each unitary transformation acting on a many-qubit system can be decomposed into gates acting on a single qubit and a single gate acting on two qubits, for instance the CNOT gate.

A large number of different proposals to build real quantum computers have been put forward. They range from NMR quantum processors to cold ion traps, superconducting tunnel-junction circuits and spin in semiconductors, to name but a few. Even though in some cases elementary quantum gates have been realized and quantum algorithms with a small number of qubits demonstrated, it is too early to say what type of implementation will be the most suitable to build a scalable piece of quantum hardware. Although for some computational problems the quantum computer is more powerful than the classical computer, still we need 50–1000 qubits and from thousands to millions of quantum gates to perform tasks inaccessible to the classical computer (the exact numbers depend, of course, on the specific quantum algorithm).

The technological challenge of realizing a quantum computer is very demanding: we need to be able to control the evolution of a large number of qubits for the time necessary to perform many quantum gates. Decoherence may be considered the ultimate obstacle to the practical realization of a quantum computer. Here the term decoherence denotes the decay of the quantum information stored in a quantum computer, due to the inevitable interaction of the quantum computer with the environment. Such interaction affects the performance of a quantum computer, introducing errors into the computation. Another source of errors that must be taken into account is the presence of imperfections in the quantum-computer hardware. Even though quantum error-correcting codes exist, a necessary requirement for a successful correction procedure is that one can implement many quantum gates inside the decoherence time scale. Here "many" means 10^3–10^4, the exact value depending on the kind of error. It is very hard to fulfil this requirement in complex many-qubit quantum systems.

The following question then arises: is it possible to build a useful quantum computer that could outperform existing classical computers in important computational tasks? And, if so, when? Besides the problem of decoherence, we should also remark on the difficulty of finding new and efficient quantum algorithms. We know that the integer-factoring problem can be solved efficiently on a quantum computer, but we do not know the answer to the following fundamental question: What class of problems could be simulated efficiently on a quantum computer? Quantum computers open up fascinating prospects, but it does not seem likely that they will become a reality with practical applications in a few years. How long might it take to develop the required technology? Even though unexpected technological breakthroughs are, in principle, always possible, one should remember the enormous effort that was necessary in order to develop the technology of classical computers.

Nevertheless, even the first, modest, demonstrative experiments are remarkable, as they allow for testing the theoretical principles of quantum mechanics. Since quantum mechanics is a particularly counter-intuitive theory, we should at the very least expect that experiments and theoretical studies on quantum computation will provide us with a better understanding of quantum mechanics. Moreover, such research stimulates the control of individual quantum systems (atoms, electrons, photons *etc.*). We stress that this is not a mere laboratory curiosity, but has interesting technological applications. For instance, it is now possible to realize single-ion clocks that are more precise than standard atomic clocks. In a sense we may say that quantum computation rationalizes the efforts of the various experiments that manipulate individual quantum systems.

Another important research direction concerns the (secure) transmission of information. In this case, quantum mechanics allows us to perform not only faster operations but also operations *inaccessible* to classical means. Entanglement is at the heart of many quantum-information protocols. It is the most spectacular and counter-intuitive manifestation of quantum mechanics, observed in composite quantum systems: it signifies the existence of non-local correlations between measurements performed on well-separated particles. After two classical systems have interacted, they are in well-defined individual states. In contrast, after two quantum particles have interacted, in general, they can no longer be described independently of each other. There will be purely quantum correlations between two such particles, independently of their spatial separation. This is the content of the celebrated EPR paradox, a *Gedanken* experiment proposed by Einstein, Podolsky and Rosen in 1935. These authors showed that quantum theory leads to a contradiction, provided that we accept the two, seemingly natural, principles of realism and locality. The reality principle states that, if we can predict with certainty the value of a physical quantity, then this value has physical reality, independently of our observation. The locality principle states that, if two systems are causally disconnected, the results of any measurement performed on one system cannot influence the result of a measurement performed on the second system. In other words, information cannot travel faster than the speed of light.

In 1964 Bell proved that this point of view (known as local realism) leads to predictions, Bell's inequalities, that are in contrast with quantum theory. Aspect's experiments (1982), performed with pairs of entangled photons, exhibited an unambiguous violation of a Bell's inequality by tens of standard deviations and an impressive agreement with quantum mechanics. These experiments also showed that it is possible to perform laboratory investigations on the more fundamental, non-intuitive aspects of quantum theory. More recently, other experiments have come closer to the requirements of the ideal EPR scheme. More generally, thanks to the development and increasing precision of experimental techniques, *Gedanken* experiments of the past become present-day real experiments.

The profound significance of Bell's inequalities and Aspect's experiments lies far beyond that of a mere consistency test of quantum mechanics. These results show that entanglement is a fundamentally new resource, beyond the realm of classical physics, and that it is possible to experimentally manipulate entangled states.

Quantum entanglement is central to many quantum-communication protocols. Of particular importance are *quantum dense coding*, which permits transmission of two bits of classical information through the manipulation of only one of two entangled qubits, and *quantum teleportation*, which allows the transfer of the state of one quantum system to another over an arbitrary distance. In recent experiments, based on photon pairs, entanglement has been distributed with the use of optical-fibre links, over distances of up to 10 kilometres. The long-distance free-space distribution of entanglement has also been recently demonstrated, with the two receivers of the entangled photons being separated by 600 metres. It is important to point out that the turbulence encountered along such an optical path is comparable to the effective turbulence in an earth-to-satellite transmission. Therefore, one may expect that in the near future it will become possible to distribute entanglement between receivers located very far apart (in two different continents, say) using satellite-based links.

Quantum mechanics also provides a unique contribution to cryptography: it enables two communicating parties to detect whether the transmitted message has been intercepted by an eavesdropper. This is not possible in the realm of classical physics as it is always possible, in principle, to copy classical information without changing the original message. In contrast, in quantum mechanics the measurement process, in general, disturbs the system for fundamental reasons. Put plainly, this is a consequence of the Heisenberg uncertainty principle. Experimental advances in the field of quantum cryptography are impressive and quantum-cryptographic protocols have been demonstrated, using optical fibres, over distances of a few tens of kilometres at rates of the order of a thousand bits per second. Furthermore, free-space quantum cryptography has been demonstrated over distances up to several kilometres. In the near future, therefore, quantum cryptography could well be the first quantum-information protocol to find commercial applications.

To conclude this introduction, let us quote Schrödinger [*Brit. J. Phil. Sci.*, **3**, 233 (1952)]: "*We never experiment with just one electron or atom or (small) molecule. In thought-experiments we sometimes assume that we do; this invariably entails ridiculous consequences ... we are not experimenting with single particles, any more than we can raise Ichthyosauria in the zoo.*" It is absolutely remarkable that only fifty years later experiments on single electrons, atoms and molecules are routinely performed in laboratories all over the world.

A guide to the bibliography

We shall conclude each chapter with a short guide to the bibliography. Our aim is to give general references that might be used by the reader as an entry point for a more in-depth analysis of the topics discussed in this book. We shall therefore often refer to review papers instead of the original articles.

General references on quantum information and computation are the lecture notes of Preskill (1998) and the books of Gruska (1999) and Nielsen and Chuang (2000). Introductory level texts include Williams and Clearwater (1997), Pittenger (2000) and Hirvensalo (2001). Useful lecture notes have been prepared by Aharonov (2001), Vazirani (2002) and Mermin (2003). Mathematical aspects of quantum computation are discussed in Brylinski and Chen (2002). Interesting collections of review papers are to be found in Lo *et al.* (1998), Alber *et al.* (2001), Lomonaco (2002) and Bouwmeester *et al.* (2000). This last text is particularly interesting from the point of view of experimental implementations.

Useful review papers of quantum computation and information are those of Steane (1998) and Galindo and Martin-Delgado (2002). The basic concepts of quantum computation are discussed in Ekert *et al.* (2001). A very readable review article of quantum information and computation is due to Bennett and DiVincenzo (2000).

A bibliographic guide containing more than 8000 items (updated as of June 2003) on the foundations of quantum mechanics and quantum information can be found in Cabello (2000–2003).

Chapter 1

Introduction to Classical Computation

This chapter introduces the basic concepts of computer science that are necessary for an understanding of quantum computation and information. We discuss the Turing machine, the fundamental model of computation since it formalizes the intuitive notion of an algorithm: if there exists an algorithm to solve a given problem, then this algorithm can be run on a Turing machine. We then introduce the circuit model of computation, which is equivalent to the Turing machine model but is nearer to real computers. In this model, the information is carried by wires and a small set of elementary logical operations (gates) allows implementation of any complex computation. It is important to find the minimum resources (computer memory, time and energy) required to solve a given problem with the best possible algorithm. This is the task of computational complexity, for which we provide a quick glance at the key concepts. Finally, we examine the energy resources necessary to perform computations. Here we discuss the relation between energy and information, which was explained by Landauer and Bennett in their solution of Maxwell's demon paradox. In particular, Landauer's principle sets the minimum energy requirements for irreversible computation. On the other hand, it turns out that it is, in principle, possible to perform any complex computation by means of reversible gates, without energy dissipation. A concrete model of reversible computation, the so-called billiard-ball computer, is briefly discussed.

1.1 The Turing machine

An *algorithm* is a set of instructions for solving a given problem. Examples of algorithms are those learnt at primary schools for adding and multiplying two integer numbers. Such algorithms always give the correct result when applied to any pair of integer numbers.

The *Turing machine*, introduced by the mathematician Alan Turing in the 1930's, provides a precise mathematical formulation of the intuitive concept of algorithm. This machine contains the essential elements (memory, control unit and read/write unit) on which any modern computer is based. Turing's work was stimulated by an intense debate at that time regarding the following question: for which class or classes of problems is it possible to find an algorithm? This debate was motivated by a profound question raised by David Hilbert at the beginning of the twentieth century. Hilbert asked whether or not an algorithm might exist that could, in principle, be used to solve all mathematical problems. Hilbert thought (erroneously, as we shall see in this section) that the answer to his question was positive.

A closely related problem is the following: given a logical system defined by an ensemble of axioms and rules, can all possible propositions be proved, at least in principle, to be either true or false? At the beginning of twentieth century it was widely believed that the answer to this question was also positive. (Of course, the question does not address the problem that, in practice, it may be extremely difficult to prove whether a proposition is true or false). Contrary to this belief, in the 1930's Kurt Gödel proved a theorem stating that there exist mathematical propositions of any given logical system that are *undecidable*, meaning that they can neither be proved nor disproved using axioms and rules inside the same logical system. This does not exclude that we can enlarge the system, introducing new axioms and rules, and thus decide whether a given proposition is true or false. However, it will also be possible to find undecidable propositions in this new system. Thus, it turns out that logical systems are intrinsically *incomplete*. Notice that Gödel's theorem also sets limits on the possibilities of a computer: it cannot answer all questions on arithmetics.

The main elements of a Turing machine are illustrated in Fig. 1.1. The general idea is that the machine performs a computation as a "human computer" would. Such a human computer is capable of storing only a limited amount of information in his brain, but has at his disposal an (ideally) unlimited amount of paper for reading and writing operations. Likewise, the Turing machine contains the following three main elements:

1. A *tape*, which is infinite and divided into cells. Each cell holds only one letter a_i from a finite alphabet $\{a_1, a_2, \ldots, a_k\}$ or is blank. Except for a finite number of cells, all the other cells are blank.

2. A *control unit*, which has a finite number of states, $\{s_1, s_2, \ldots, s_l, H\}$,

where H is a special state, known as the halting state: if the state of the control unit becomes H, then the computation terminates.

3. A *read/write head*, which addresses a single cell of the tape. It reads and (over)writes or erases a letter in this cell, after which, the head moves one cell to the left or to the right.

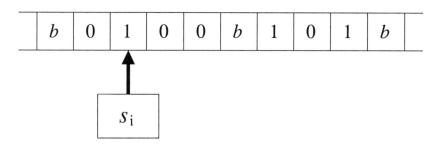

Fig. 1.1   Schematic drawing of a Turing machine. The symbol b denotes a blank cell.

The working of a Turing machine is governed by a *program*, which is simply a finite set of instructions. Each instruction governs one step of the Turing machine and induces the following sequence of operations:

(i)   the transition of the control unit from the state s to the state $\bar{s}$,

(ii) the transition of the cell addressed by the read/write head from the letter a to the letter $\bar{a}$,

(iii) the displacement of the read/write head one cell left or right.

Therefore, an instruction in the Turing machine is defined by three functions f_S, f_A and f_D, defined as follows:

$$\bar{s} = f_S(s, a), \qquad (1.1a)$$
$$\bar{a} = f_A(s, a), \qquad (1.1b)$$
$$d = f_D(s, a), \qquad (1.1c)$$

where d indicates the displacement of the head to the left $(d = l)$ or to the right $(d = r)$. In short, the functions f_S, f_A and f_D define the mapping

$$(s, a) \rightarrow (\bar{s}, \bar{a}, d). \qquad (1.2)$$

1.1.1   *Addition on a Turing machine*

Let us now describe a concrete example: a Turing machine performing the
addition of two integers. For the sake of simplicity, we write the integer
numbers in the unary representation: an integer N is written as a sequence
of N 1's, that is, $1 = 1$, $2 = 11$, $3 = 111$, $4 = 1111$ and so on. As
an example, we compute the sum $2 + 3$. Our Turing machine needs five
internal states $\{s_1, s_2, s_3, s_4, H\}$ and a unary alphabet, namely, the single
letter 1. We denote by b the blank cells of the tape. The initial condition
of the machine is shown in Fig. 1.2: the initial state is s_1, the head points
to a well-defined cell and the numbers to be added, $2 = 11$ and $3 = 111$,
are written on the tape, separated by a blank.

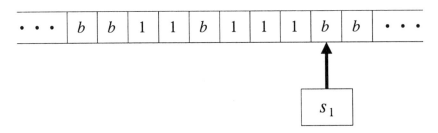

Fig. 1.2   Initial conditions of a Turing machine for computing the sum $2 + 3$.

The program for computing the sum of two integer numbers is shown in
Table 1.1. The program has a total number of six lines. The internal state

Table 1.1   The algorithm for computing the
sum of two integers on a Turing machine.

s	a	$\bar{s}$	$\bar{a}$	d
s_1	b	s_2	b	l
s_2	b	s_3	b	l
s_2	1	s_2	1	l
s_3	b	H	b	0
s_3	1	s_4	b	r
s_4	b	s_2	1	l

s of the machine and the letter a being read on the tape determine which
program line is executed. The last three columns of Table 1.1 denote the
new state $\bar{s}$, the letter $\bar{a}$ overwritten on the tape and the left/right direction
($d = l$ or $d = r$) of the read/write head motion. Note that in the fourth

line of the program $d = 0$ since the machine halts and the head moves
no further. It is easy to check that, if we start from the initial conditions
of Fig. 1.2 and run the program of Table 1.1, the machine halts in the
configuration depicted in Fig. 1.3 and we can read on the tape the result
of the sum, $2 + 3 = 5$. It is also easy to convince ourselves that the same
program can compute the sum of two arbitrary integers m and n, provided
that the initial conditions are set as in Fig. 1.4.

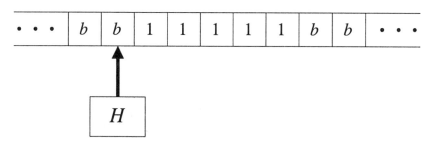

Fig. 1.3   A Turing machine after computation of the sum $2 + 3$. The machine started
from the initial conditions of Fig. 1.2 and implemented the program of Table 1.1.

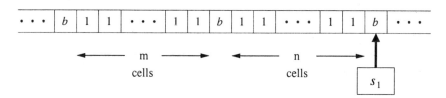

Fig. 1.4   The initial conditions of a Turing machine for computing the sum $m + n$ of
two generic integers.

1.1.2   *The Church–Turing thesis*

It turns out that Turing machines are capable of solving very complex
problems. As far as we know, they can be used to simulate any operation
carried out on a modern computer. If there exists an algorithm to compute
a function, then the computation can be performed by a Turing machine.
This idea was formalized independently by Church and Turing:

The Church–Turing thesis: *The class of all functions computable by
a Turing machine is equivalent to the class of all functions computable by
means of an algorithm.*

This statement provides a rigorous mathematical definition of the intuitive concept of "function computable by an algorithm": a function is computable if and only if it can be computed by a Turing machine. The thesis, formulated in 1936, has never been disproved since we do not know of any algorithm that computes a function not computable by a Turing machine. Indeed, much evidence has been gathered in favour of the Church–Turing thesis.

1.1.3   *The universal Turing machine*

The universal Turing machine U is a single machine that encompasses all Turing machines; that is, it is capable of computing any algorithm. A Turing machine T running a given program, on the basis of the input x written on tape, produces some output $T(x)$ and then halts. The universal Turing machine can simulate any Turing machine T, provided that on the tape of the universal Turing machine we specify the description of the machine T. It can be shown that an integer number n_T may be uniquely associated with the corresponding machine T. This number is known as the *Turing number* associated with the machine. Therefore, if we give the description n_T of T and x on input, the universal Turing machine U produces the output $U(n_T, x) = T(x)$. It is important to stress that in the universal Turing machine the (finite) set of internal states $\{s_i\}$ and the program are fixed once and for all. Thus, we can run any computation by simply changing the initial state of the tape.

1.1.4   *The probabilistic Turing machine*

A probabilistic Turing machine is characterized by the fact that the mapping $(s, a) \rightarrow (\bar{s}, \bar{a}, d)$ is probabilistic. This means that there exist *coin-tossing states* in which the machine tosses a coin to decide the output. The coin lands heads up with probability p and tails up with probability $1 - p$. In the first case the new internal state of the machine is given by $\bar{s} = \bar{s}_h$ while in the latter we have $\bar{s} = \bar{s}_t$. A probabilistic Turing machine may be more powerful than a deterministic Turing machine, in that it can solve many computational problems faster. An example illustrating the usefulness of probabilistic algorithms will be discussed in Sec. 1.3. However, we should note that the probabilistic Turing machine does not enlarge the class of functions computable by a deterministic Turing machine. Indeed, a deterministic Turing machine can always simulate a probabilistic Turing

(so the Turing machine is enumerable)

machine by exploring, one after another, all possible paths corresponding to different values of the coin tosses.

in the sense that it provides all possible outputs as the probabilistic Tm

1.1.5 * The halting problem

Let us now consider the following problem: will some given Turing machine T eventually halt for some given input x? The question is quite natural: the machine could either end up in the internal state H and stop after some finite time or loop infinitely without ever reaching the state H. Turing demonstrated that there exists no algorithm capable of solving this problem, known as the halting problem. An instance of this problem is the following: will a given machine T attain the halt state H after input of its own Turing number n_T? In other words, is there an algorithm (or Turing machine) A whose output $A(n_T)$ tells us whether or not some Turing machine T eventually halts on input of n_T?

Let us assume that such an algorithm exists. In other words, if machine T halts for input n_T, then for the same input A writes "*yes*" and halts, otherwise A writes "*no*" and halts. In the following, we shall prove that such a machine A cannot exist. Let us consider another machine B, defined as follows: if A writes "*yes*" for some input n_T then B does not halt, if instead A writes "*no*" then B does halt. If A exists, then B exists as well. Therefore, for any n_T, $B(n_T)$ halts if and only if $T(n_T)$ does not halt. We now consider the case in which the input of the machine B is its own Turing number n_B. Therefore, $B(n_B)$ halts if and only if $B(n_B)$ does not halt. This is a contradiction and therefore the machine A cannot exist.

To understand the logical basis of this proof by *reductio ad absurdum*, consider the following paradoxical statement: "This sentence is false." While it does not violate any grammatical law, that is, its construction is perfectly legitimate, there is no possible answer to the question "Is the statement true or not?" The above problem is equivalent to asking a computer to provide just such answer. However, it should also be remarked that the *practical* stumbling block to such an algorithm is clearly the difficulty in demonstrating the infinite-loop condition.

1.2 The circuit model of computation

In terms of computational power, the circuit model of computation is equivalent to the Turing machine model discussed in the previous section but is nearer to a real computer. Let us first introduce the *bit*, the elementary unit of classical information. A bit is defined as a two-valued or binary variable, whose values are typically written as the binary digits 0 and 1. A circuit is made of *wires* and *gates*; each wire carries one bit of information since it may take the value 0 or 1. As we shall see below, the gates perform logic operations on these bits. The classical computer is a *digital* device since the information enters the computer as a sequence of 0's and 1's and the output of any computation is again a sequence of 0's and 1's. For instance, an integer number $N < 2^n$ is stored as a binary sequence of 0's and 1's as follows:

$$N = \sum_{k=0}^{n-1} a_k 2^k \,, \tag{1.3}$$

where the value of each binary digit a_k may be equal to 0 or to 1. We may write equivalently

$$N = a_{n-1} a_{n-2} \ldots a_1 a_0 \,. \tag{1.4}$$

For instance, we have $3 = 11$, $4 = 100$, $5 = 101$ and $49 = 110001$. We may also write binary fractions, using the notation $\frac{1}{2} = 0.1$, $\frac{1}{4} = 0.01$, $\frac{1}{8} = 0.001$ and so on. Let us write the binary codes of a few non-integer numbers: $5.5 = 101.1$, $5.25 = 101.01$ and $5.125 = 101.001$. It is clear that any real number may be approximated to any desired accuracy by a binary fraction.

The advantage of the binary notation is that binary numbers are well suited to being stored in electrical devices since only two possible values need be set: computers use high and low voltage values or switches with only two positions (*on* and *off*) to load one bit of information. For instance, in Fig. 1.5 we show the sequence of voltages required to load the integer number $N = 49$.

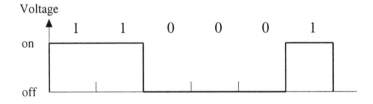

Fig. 1.5   The sequence of voltages representing the integer $N = 49$.

1.2.1 *Binary arithmetics*

The arithmetical rules also turn out to be much simpler in the binary representation. As an example, in Table 1.2 we show the binary addition table, where $s = a \oplus b$ is the addition, modulo two, of the two bits a and b while c is the carry over.

Table 1.2   The binary addition table.

a	b	s	c
0	0	0	0
0	1	1	0
1	0	1	0
1	1	0	1

The following examples should help clarify the procedure for computing the sum and product of two numbers written in their binary representations. Decimal addition and multiplication are also shown to provide a more familiar comparison.

```
                              BINARY              DECIMAL

ADDITION                 1 1 1 0 1                   2 9
                         1 0 1 0 1                   2 1
                     ─────────────                 ─────
                       1 1 0 0 1 0                   5 0


MULTIPLICATION           1 1 1 0 1                   2 9
                         1 0 1 0 1                   2 1
                     ─────────────                 ─────
                         1 1 1 0 1                   2 9
                       1 1 1 0 1                   5 8
                     1 1 1 0 1                   ─────
                   ─────────────────              6 0 9
                   1 0 0 1 1 0 0 0 0 1
```

1.2.2 *Elementary logic gates*

In any computation, we must provide an n-bit input to recover an l-bit output. Namely, we must compute a logical function of the form

$$f : \{0,1\}^n \rightarrow \{0,1\}^l. \tag{1.5}$$

As we shall show later in this section, the evaluation of any such function may be decomposed into a sequence of elementary logical operations. First of all, we introduce a few logic gates that are useful for computation.

Fig. 1.6 shows a trivial one-bit gate, the identity gate: the value of the output bit is simply equal to the value of the input bit. The simplest non-

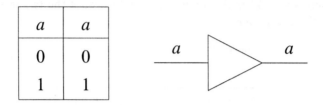

a	a
0	0
1	1

Fig. 1.6   The truth table and circuit representation for the identity gate.

trivial gate is the NOT gate, which acts on a single bit and flips its value: if the input bit is 0, the output bit is set to 1 and vice versa. In binary arithmetics,

$$\bar{a} = 1 - a, \qquad (1.6)$$

where $\bar{a}$ denotes NOT a. The circuit representation and the truth table for the NOT gate are shown in Fig. 1.7.

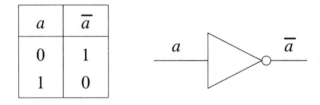

a	$\bar{a}$
0	1
1	0

Fig. 1.7   The truth table and circuit representation for the NOT gate.

We next introduce a list of two-bit logic gates useful for computation. These gates have two input bits and one output bit and are therefore binary functions $f : \{0,1\}^2 \to \{0,1\}$.

(i) the AND ($\wedge$) gate (see Fig. 1.8): produces output 1 if and only if both input bits are set to 1. In binary arithmetics,

$$a \wedge b = ab. \qquad (1.7)$$

a	b	$a \wedge b$
0	0	0
0	1	0
1	0	0
1	1	1

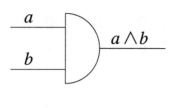

Fig. 1.8   The truth table and circuit representation for the AND gate.

(ii) the OR ($\vee$) gate (see Fig. 1.9): produces output 1 if and only if at least one of the input bits is set to 1. In binary arithmetics,

$$a \vee b = a + b - ab. \tag{1.8}$$

a	b	$a \vee b$
0	0	0
0	1	1
1	0	1
1	1	1

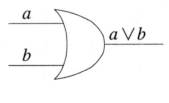

Fig. 1.9   The truth table and circuit representation for the OR gate.

(iii) the XOR ($\oplus$) gate (see Fig. 1.10): produces output 1 if only one of the input bits is set to 1, otherwise the output is 0. The XOR (also known as the exclusive OR) gate outputs the sum, modulo 2, of the inputs:

$$a \oplus b = a + b \quad (\mathrm{mod}\, 2). \tag{1.9}$$

(iv) the NAND ($\uparrow$) gate (see Fig. 1.11): produces output zero if and only if both inputs are set to one. It is obtained by the application of a NOT

a	b	$a \oplus b$
0	0	0
0	1	1
1	0	1
1	1	0

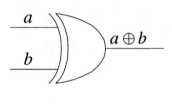

Fig. 1.10   The truth table and circuit representation for the XOR gate.

gate to the output of an AND gate:

$$a \uparrow b = \overline{a \wedge b} = \overline{ab} = 1 - ab \,. \tag{1.10}$$

a	b	$a \uparrow b$
0	0	1
0	1	1
1	0	1
1	1	0

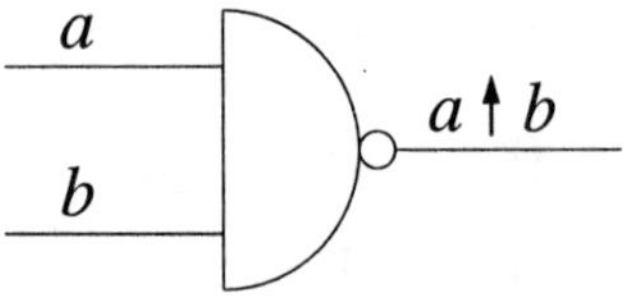

Fig. 1.11   The truth table and circuit representation for the NAND gate.

(v) the NOR ($\downarrow$) gate (see Fig. 1.12): produces output 1 if and only if both
     inputs are set to zero. It is obtained by the application of a NOT gate
     to the output of an OR gate:

$$a \downarrow b = \overline{a \vee b} = \overline{a + b - ab} = 1 - a - b + ab \,. \tag{1.11}$$

Other important gates are the FANOUT (also known as COPY) gate,
which takes one bit into two bits:

$$\text{COPY} : a \rightarrow (a, a) \,, \tag{1.12}$$

a	b	$a \downarrow b$
0	0	1
0	1	0
1	0	0
1	1	0

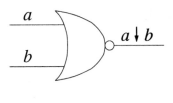

Fig. 1.12   The truth table and circuit representation for the NOR gate.

and the CROSSOVER (or SWAP), which interchanges the values of two bits:

$$\text{CROSSOVER} : (a, b) \rightarrow (b, a). \qquad (1.13)$$

The circuit representations for these two gates are shown in Fig. 1.13.

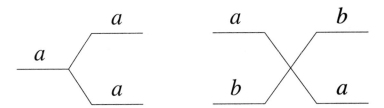

Fig. 1.13   Circuit representations for the FANOUT (COPY) gate (left) and the CROSSOVER (SWAP) gate (right).

The elementary gates described above may be put together to implement any complex computation. As an example, we shall construct a circuit to perform the summation s of two integer numbers a and b using a sequence of AND, OR, XOR and FANOUT gates. We can compute the sum $s = a + b$ bit-by-bit. Given the binary representations $a = (a_n, a_{n-1}, \ldots, a_1, a_0)$ and $b = (b_n, b_{n-1}, \ldots, b_1, b_0)$, the i-th bit of the sum is

$$s_i = a_i \oplus b_i \oplus c_i \qquad (\text{mod } 2), \qquad (1.14)$$

where c_i is the carry over from the sum $a_{i-1} \oplus b_{i-1} \oplus c_{i-1}$. We denote c_{i+1} the carry over of the sum (1.14): it is set to 1 if two or more of the input bits a_i, b_i and c_i are 1 and 0 otherwise. It is easy to check that the circuit of Fig. 1.14 takes as input a_i, b_i and c_i and outputs s_i and c_{i+1}.

$\Bigg\{ \quad s_i = a_i + b_i + c_i \mod 2$

$c_{i+1} = a_i b_i + b_i c_i + c_i c_i \mod 2$

[handwritten annotations: instead of computing each term separately the diagram below computes
$(a_i + b_i)c_i + a_ib_i$
$(a_i + b_i) + a_i$
$\Rightarrow a_i + b_i$ used repeatedly]*

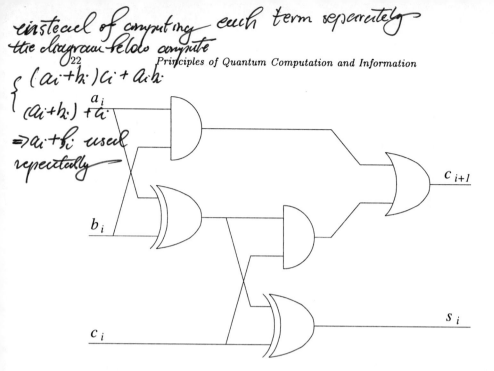

Fig. 1.14   A circuit for computing the sum $s_i = a_i \oplus b_i \oplus c_i$ and the carry c_{i+1}. The bifurcating wires are achieved by means of FANOUT gates.

It is important to note that the elementary gates introduced above are not all independent. For instance, AND, OR and NOT are related by De Morgan's identities:

$$\overline{a \wedge b} = \bar{a} \vee \bar{b}, \tag{1.15a}$$

$$\overline{a \vee b} = \bar{a} \wedge \bar{b}. \tag{1.15b}$$

It is also easy to check that the XOR gate can be constructed by means of the AND, OR and NOT gates as follows:

$$a\,\text{XOR}\,b = (a\,\text{OR}\,b)\,\text{AND}\,\big((\text{NOT}\,a)\,\text{OR}\,(\text{NOT}\,b)\big). \tag{1.16}$$

1.2.3   Universal classical computation

Universal gates: *Any function*

$$f : \{0,1\}^n \to \{0,1\}^m \tag{1.17}$$

can be constructed from the elementary gates AND, OR, NOT and FANOUT. Therefore, we say that these constitute a universal set of gates for classical computation.

Proof. The m-bit function (1.17) is equivalent to m one-bit (Boolean) functions

$$f_i : \{0,1\}^n \to \{0,1\}, \qquad (i = 1, 2, \ldots, m), \qquad (1.18)$$

where $f = (f_1, f_2, \ldots, f_m)$. One way to compute these Boolean functions $f_i(a)$, $a = (a_{n-1}, a_{n-2}, \ldots, a_1, a_0)$, is to consider its *minterms* $f_i^{(l)}(a)$, defined, for each $a^{(l)}$ such that $f_i(a^{(l)}) = 1$, as

$$f_i^{(l)}(a) = \begin{cases} 1 & \text{if } a = a^{(l)}, \\ 0 & \text{otherwise.} \end{cases} \qquad (1.19)$$

Then the function $f_i(a)$ reads as follows:

$$f_i(a) = f_i^{(1)}(a) \vee f_i^{(2)}(a) \vee \cdots \vee f_i^{(k)}(a), \qquad (1.20)$$

where $f_i(a)$ is the logical OR of all k minterms (with $0 \leq k \leq 2^n - 1$). It is therefore sufficient to compute the minterms and to perform the OR gates in order to obtain $f_i(a)$. We note that the decomposition (1.20) also requires the implementation of FANOUT gates. We need k copies of the input a, since each minterm must act on it.

The evaluation of $f_i^{(l)}$ may be performed as follows. If, for instance, $a^{(l)} = 110100 \ldots 001$, we have

$$f_i^{(l)}(a) = a_{n-1} \wedge a_{n-2} \wedge \overline{a}_{n-3} \wedge a_{n-4} \wedge \overline{a}_{n-5} \wedge \overline{a}_{n-6} \wedge \ldots \wedge \overline{a}_2 \wedge \overline{a}_1 \wedge a_0. \quad (1.21)$$

Thus, $f_i^{(l)}(a) = 1$ if and only if $a = a^{(l)}$. This completes our proof: we have constructed a generic function $f(a)$ from the elementary logic gates AND, OR, NOT and FANOUT. $\qquad \square$

As an illustration of the above procedure, we consider the Boolean function $f(a)$, where $a = (a_2, a_1, a_0)$, defined as follows: $f(a) = 1$ if $a = a^{(1)} = 1$ ($a_2 = 0$, $a_1 = 0$, $a_0 = 1$) or if $a = a^{(2)} = 3$ ($a_2 = 0$, $a_1 = 1$, $a_0 = 1$) or if $a = a^{(3)} = 6$ ($a_2 = 1$, $a_1 = 1$, $a_0 = 0$) and $f(a) = 0$ otherwise. The minterms of $f(a)$ are $f^{(1)}(a)$, $f^{(2)}(a)$ and $f^{(3)}(a)$, which are equal to one if and only if $a = a^{(1)}$, $a = a^{(2)}$ and $a = a^{(3)}$, respectively. We have $f(a) = f^{(1)}(a) \vee f^{(2)}(a) \vee f^{(3)}(a)$, where $f^{(1)}(a) = \overline{a}_2 \wedge \overline{a}_1 \wedge a_0$, $f^{(2)}(a) = \overline{a}_2 \wedge a_1 \wedge a_0$ and $f^{(3)}(a) = a_2 \wedge a_1 \wedge \overline{a}_0$.

Actually, it is even possible to reduce the number of elementary operations. It turns out, for example, that NAND and FANOUT are a smaller universal set. Indeed, we have already seen that OR can be obtained from NOT and AND by means of De Morgan's identities. It is also easy to obtain NOT from NAND and FANOUT:

$$a \uparrow a = \overline{a \wedge a} = 1 - a^2 = 1 - a = \overline{a}. \qquad (1.22)$$

Exercise 1.1 Construct AND and OR from NAND and FANOUT.

In computers the NAND gate is usually implemented via transistors, as shown in Fig. 1.15. A bit is set to 1 if the voltage is positive and to

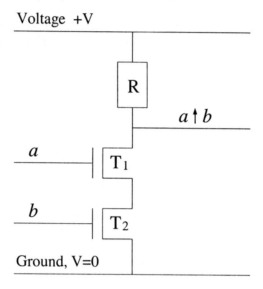

Fig. 1.15   The electrical circuit for a NAND gate. R denotes a resistor, T_1 and T_2 two transistors.

0 if the voltage is zero. It is easy to verify that the output is a NAND b. Indeed, the current flows through the transistors if and only if both inputs have positive voltage ($a = b = 1$). In this case, the output has zero voltage. If at least one of the inputs has zero voltage, there is no current flow and therefore the output has positive voltage.

1.3   Computational complexity

To solve any given problem, a certain amount of resources is necessary. For instance, to run an algorithm on a computer we need *space* (that is, memory), *time* and *energy*. Computational complexity is the study of the resources required to solve computational problems. Sometimes it is immediately obvious if one problem is easier to solve than another. For instance, we know that it is easier to add two numbers than to multiply them. In other cases, it may be very difficult to evaluate and measure the complexity of a problem. This is a particularly important objective, which affects many fields, from computer science to mathematics, physics, biology, medicine, economics and even the social sciences. An important task of computational complexity is to find the *minimum* resources required to solve a given problem with the best possible algorithm.

Let us consider a simple example. As noted above, intuitively it is easier to add two numbers than to multiply them. This statement is based on two algorithms learnt at primary school: the addition of two n-digit integer numbers requires a number of steps that grows linearly with n, that is, the time necessary to execute the algorithm is $t_a = \alpha n$. The number of steps required to compute the multiplication of these two numbers is instead proportional to the square of n: $t_m = \beta n^2$. Therefore, one might be tempted to conclude that multiplication is more complex than addition. However, such a conclusion would be based on particular algorithms for computing addition and multiplication: those learnt at primary school. Could different algorithms lead us to different conclusions? It is clear that the addition of two numbers cannot be performed in a number of steps smaller than n: we must at least read the two n-digit input numbers. Therefore, we may conclude that the complexity of addition is $O(n)$ (given two functions $f(n)$ and $g(n)$, we say that $f = O(g)$ if, for $n \to \infty$, $c_1 \leq |f(n)/g(n)| \leq c_2$, with $0 \leq c_1 \leq c_2 < \infty$). On the other hand, in 1971 Schönhage and Strassen discovered an algorithm, based on the fast Fourier transform, that requires $O(n \log n \log \log n)$ steps to carry out the multiplication of two n-digit numbers on a Turing machine. Is there a better algorithm to compute multiplication? If not, we should conclude that the complexity of multiplication is $O(n \log n \log \log n)$ and that addition is easier than multiplication. However, we cannot exclude that better algorithms for computing multiplication might exist.

The main distinction made in the theory of computational complexity is between problems that can be solved using *polynomial* versus *exponential* resources. More precisely, if n denotes the *input size*, that is, the number of bits required to specify the input, we may divide solvable problems into two classes:

1. Problems that can be solved using resources that are bounded by a polynomial in n. We say that these problems can be solved *efficiently*, or that they are *easy*, *tractable* or *feasible*. Addition and multiplication belong to this class.
2. Problems requiring resources that are *superpolynomial* (*i.e.*, which grow faster than any polynomial in n). These problems are considered as *difficult*, *intractable or unfeasible*. For instance, it is believed (though

not proven) that the problem of finding the prime factors of an integer number is in this class. That is, the best known algorithm for solving this problem is superpolynomial in the input size n, but we cannot exclude that a polynomial algorithm might exist.

Comments

(i) An example may help us clarify the difficulty of superpolynomial problems: the best known algorithm for the factorization of an integer N, the number field sieve, requires $\exp(O(n^{1/3}(\log n)^{2/3}))$ operations, where $n = \log N$ is the input size. Thus, the factorization of a number 250 digits long would take about 10 million years on a 200-MIPS (million of instructions per second) computer (see Hughes, 1998). Therefore, we may conclude that the problem is in practice impossible to solve with existing algorithms and any conceivable technological progress.

(ii) It is clear that also a polynomial algorithm scaling as n^α, with $\alpha \gg 1$, say $\alpha = 1000$, can hardly be regarded as easy. However, it is in practice very unusual to encounter useful algorithms with $\alpha \gg 1$. In addition, there is a more fundamental reason to base the theory of computational complexity on the distinction between polynomial and exponential algorithms. Indeed, according to the strong Church–Turing thesis, this classification is *robust* when the model of computation is changed.

The strong Church–Turing thesis: *A probabilistic Turing machine can simulate any model of computation with at most a polynomial increase in the number of elementary operations required.*

This thesis states that, if a problem cannot be solved with polynomial resources on a probabilistic Turing machine, it has no efficient solution on any other machine. Any model of computation is at best polynomially equivalent to the probabilistic Turing machine model.

In this connection it is interesting to note that quantum computers challenge the strong Church–Turing thesis. Indeed, as will be shown in Chap. 3, there exists an algorithm, discovered by Peter Shor, that solves the integer factorization problem on a quantum computer with polynomial resources. As discussed above, we do not know of any algorithm that solves this problem polynomially on a classical computer. And indeed, if such an algorithm does not exist, then we should conclude that the quantum model of computation is more powerful than the probabilistic Turing machine model and the strong Church–Turing thesis should be rejected.

1.3.1 *Complexity classes*

We say that a problem belongs to the computational class **P** if it can be solved in *polynomial time*, namely, in a number of steps that is polynomial in the input size. The computational class **NP**, instead, is defined as the class of problems whose solution can be verified in polynomial time. It is clear that **P** is a subset of **NP**, namely, **P** ⊆ **NP**. It is a fundamental open problem of mathematics and computer science whether there exist problems in **NP** that are not in **P**. It is conjectured, though not proven, that **P** ≠ **NP**. If this were the case, there would be problems hard to solve but whose solution could be easily checked. For instance, the integer-factoring problem is in the class **NP**, since it is easy to check if a number m is a prime factor of an integer N, but we do not know of any algorithm that efficiently computes the prime factors of N (on a classical computer). Therefore, it is conjectured that the integer-factoring problem does not belong to the class **P**.

We say that a problem in **NP** is **NP**-complete (**NPC**) if any problem in **NP** is *polynomially reducible* to it. This means that, given an **NPC** problem, for any problem in **NP** there is a mapping that can be computed with polynomial resources and that maps it to the **NPC** problem. Therefore, if an algorithm capable of efficiently solving an **NPC** problem is discovered, then we should conclude that **P** = **NP**. An example of an **NPC** problem is the *Travelling Salesman Problem*: given n cities, the distances d_{jk} between them $(j, k = 1, 2, \ldots, n)$ and some length d, is there a path along which the salesman visits each city and whose length is shorter than d? We point out that some problems, notably the integer-factoring problem, are conjectured to be neither in **P** nor **NP**-complete. It has been proven that, if **P** ≠ **NP**, then there exist **NP** problems that are neither in **P** nor in **NPC**. The possible maps of **NP** problems are drawn in Fig. 1.16.

So far, we have discussed time resources. However, to run a computation space and energy resources are also important. The discussion of energy resources will be postponed to the next section.

Space (i.e., memory) resources. Space and time resources are linked. Indeed, if at any step of, say, a Turing machine we use a new memory cell, then space and time resources scale equivalently. However, there is a fundamental difference: space resources can be reused. We define **PSPACE** as the class of problems that can be solved by means of space resources that are polynomial in the input size, independently of the computation time. It is evident that **P** ⊆ **PSPACE**, since

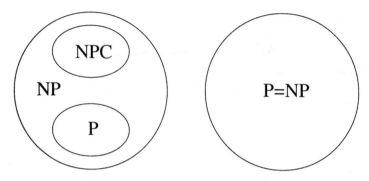

Fig. 1.16 Possible maps of **NP** problems. It is conjectured, though not proven, that the left map is the correct one.

in a polynomial time a Turing machine can explore only a polynomial number of memory cells. It is also conjectured that **P** $\neq$ **PSPACE**. Indeed, it seems reasonable that, if we have unlimited time resources and polynomial space resources, we can solve a larger class of problems than if we have polynomial time (and space) resources. However, there is no proof that there exist problems in **PSPACE** not belonging to **P**. It is easy to show that **NP** is a subset of **PSPACE**, that is, any problem in **NP** can be solved by means of polynomial space resources. Indeed, we can always try to find the solution of an **NP** problem by exhaustive search; since each possible solution can be verified in polynomial time and space for **NP** problems, we may reuse the same (polynomial) space resources to test all possible solutions. In summary, we know that **P** $\subseteq$ **NP** $\subseteq$ **PSPACE**, but we do not know if these inclusions are strict.

Finally, let us consider the case in which a probabilistic computer (such as a probabilistic Turing machine) is used to solve a decision problem, namely, a problem whose solution may only be *"yes"* or *"no"*. We say that the problem is of the **BPP** class (*bounded-error probabilistic polynomial time*) if there exists a polynomial-time algorithm such that the probability of getting the right answer is larger than $\frac{1}{2} + \delta$ for every possible input and $\delta > 0$. The Chernoff bound, discussed in the next subsection, shows that the probability of getting the right answer can be quickly amplified by running the algorithm several times and then applying majority voting. Indeed, in order to reduce the error probability below ϵ in a **BPP** problem, it is sufficient to repeat the algorithm a number of times logarithmic in $1/\epsilon$.

The following simple example demonstrates that sometimes it is convenient to relax the requirement that a solution is always correct and allow some very small error probability. Let us consider a database of N bits $j_1, ..., j_N$. Suppose that we know in advance that either they are all equal ($j_1 = ... = j_N = 0$ or $j_1 = ... = j_N = 1$) or half of them are 0 and half 1. We call the first possibility "constant" and the second "balanced". Our problem is to distinguish between these two possibilities. In the case in which the bits are all equal, we must observe $N/2 + 1$ of them to be sure of our answer to the problem. Indeed, if we observe $N/2$ bits (for instance, from j_1 to $j_{N/2}$) and they are all equal to, say, 0, we cannot exclude with certainty the balanced case: we could have $j_1 = ... = j_{N/2} = 0$ and $j_{N/2+1} = ... = j_N/2 = 1$. To solve our problem probabilistically, we toss a random number i between 1 and N and we observe j_i. This is repeated k times. If we find two different bits, then we can conclude with certainty that we are in the balanced case. If all bits are constant, we say that we are in the constant case. Of course, there is a chance that we give the wrong answer to the problem. However, the probability of obtaining the same response every time when we are in the balanced case is $1/2^{k-1}$. Therefore, we can reduce the probability of error below some level ϵ if k is such that $1/2^{k-1} < \epsilon$. This is obtained in $k = O(\log(1/\epsilon))$ bit observations, independently of N. This simple example shows that **BPP** better than **P** should be regarded as the class of problems that can be solved efficiently on a classical computer. It is evident that $\mathbf{P} \subseteq \mathbf{BPP}$, while the relation between **NP** and **BPP** is unknown.

We close this section by introducing the computational class **BQP** (*bounded-error quantum probabilistic polynomial*). We say that a decision problem is of the **BQP** class if there exists a polynomial-time quantum algorithm giving the right answer with probability larger than $\frac{1}{2} + \delta$ (with $\delta > 0$). Since the integer-factoring problem may be reduced to a decision problem, Shor's algorithm belongs to this class. Indeed, it solves the factoring problem in $O(n^2 \log n \log \log n \log(1/\epsilon))$ operations, where ϵ is the probability of error. Note that ϵ does not depend on the input size n and therefore we can take it as small as we like and still have an efficient algorithm. We stress that there is no known classical algorithm, deterministic or probabilistic, that solves this problem in a number of operations polynomial in the input size. We know that $\mathbf{P} \subseteq \mathbf{BPP} \subseteq \mathbf{BQP} \subseteq \mathbf{PSPACE}$ and it is conjectured that $\mathbf{BPP} \neq \mathbf{BQP}$, namely, that a quantum computer is more powerful than a classical computer.

1.3.2   ⋆ *The Chernoff bound*

When solving a decision problem, a probabilistic algorithm produces a non-deterministic binary output f. Let us assume, without any loss of generality, that the correct answer is $f = 1$ and the wrong answer is $f = 0$. Let us repeat the algorithm k times and then apply majority voting. At each step i $(i = 1, 2, \ldots, k)$ we obtain $f_i = 1$ with probability $p_1 > 1/2 + \delta$ and $f_i = 0$ with probability $p_0 < 1/2 - \delta$. Majority voting fails when $s_k \equiv \sum_i f_i \le k/2$. Note that the average value of s_k is larger than $k(1/2 + \delta) > k/2$. The most probable sequences $\{f_i\}$ that lead to a failure of majority voting are those in which s_k is nearest to its average value, that is, $s_k = k/2$. Such sequences occur with probability

$$
p\left(\{f_1, f_2, \ldots, f_k\};\ s_k = \frac{k}{2}\right) < \left(\frac{1}{2} - \delta\right)^{\frac{k}{2}} \left(\frac{1}{2} + \delta\right)^{\frac{k}{2}} = \frac{(1 - 4\delta^2)^{\frac{k}{2}}}{2^k}.
$$
(1.23)

Since there are 2^k possible sequences $\{f_i\}$, we may conclude that majority voting fails with probability

$$
p\left(s_k \le \frac{k}{2}\right) < 2^k \frac{(1 - 4\delta^2)^{\frac{k}{2}}}{2^k} = (1 - 4\delta^2)^{\frac{k}{2}}.
$$
(1.24)

Finally, since $1 - x \le \exp(-x)$, we obtain the *Chernoff bound*:

$$
p\left(s_k \le \frac{k}{2}\right) < \exp(-2\delta^2 k).
$$
(1.25)

Therefore, the error probability drops below ϵ after a number of runs

$$
k > \frac{1}{2\delta^2} \ln\left(\frac{1}{\epsilon}\right).
$$
(1.26)

1.4   ⋆ **Computing dynamical systems**

One of the main applications of computers is the simulation of dynamical models describing the evolution of complex systems. We refer here not only to problems of interest for physics and mathematics, but also to a much wider class of problems in different fields such as chemistry, biology, economics, medicine, engineering, social sciences, meteorology, population dynamics and so on. From the viewpoint of computational complexity, the following question naturally arises: can such complex problems be solved efficiently? More precisely, given a generic dynamical system, is it possible to find its solution at time t efficiently? That is, since the number of bits required to specify the time t is $\log t$, can we solve the problem in a number of operations polynomial in $\log t$? We shall see in this section that this is not the case for a generic dynamical system, whose evolution is typically described by non-linear equations.

1.4.1 * *Deterministic chaos*

Deterministic chaos has been one of the most significant discoveries of the last century. Let us briefly explain the meaning of the wording "determinstic chaos". A system is said to be *deterministic* when its future, as well as its past, are determined by its present state. For instance, Newton's laws of motion unambiguously determine the future (and the past) of a system, once its state at some time t_0 is assigned. On the other hand, the motion of the system can be so complex as to be indistinguishable in practice from purely *chaotic* motion. This property allows us to reconcile the determinism of physical laws and the apparent chaoticity of natural phenomena, such as turbulence, which we observe in everyday life. Hence, the term "deterministic chaos" is not self-contradictory, since a phenomenon can be both deterministic and chaotic: deterministic since it is governed by laws that fully determine its future state from initial conditions; chaotic since its motion is so complex as to be completely *unpredictable* in practice. Let us try to clarify this statement. We first consider the harmonic oscillator, namely, the simplest example of a classical solvable or so-called integrable system. Its equation of motion, $d^2x/dt^2 + \omega^2 x = 0$, can be solved analytically. The solution is $x(t) = x_0 \cos(\omega t + \phi_0)$, with x_0 and ϕ_0 the initial conditions. Given a time t, a computer can output $x(t)$ from the above solution with $O(\log t)$ operations. In contrast, for chaotic motion, as we shall see below, the number of operations required is $O(t)$. This means that, while for an integrable system the motion is predictable and computable, for a chaotic system it is not possible to predict the future "before it arrives". That is, it is not possible to describe the orbit of a chaotic system by means of an algorithm that scales better than $O(t)$: the system itself is "its own best computer".

In order to clarify this concept, let us consider a conservative system described by the Hamiltonian $H(q,p)$, where $q = (q_1, \ldots, q_n)$ and $p = (p_1, \ldots, p_n)$ denote canonical variables. Since the total energy E is a constant of motion, the system's orbit moves on the constant-energy surface, defined by the equation $H(q,p) = E$. We now make a *partition* of this surface, that is, we divide it into a finite set of non-overlapping cells and we identify each cell by means of an integer. If we have perfect knowledge of the system's orbit, we can assign, at time intervals τ, the number of the cell in which the system resides. In this way, we obtain a sequence of integers, which provides a coarse-grained description of the orbit. For a chaotic system, no regularity appears. The knowledge of the cells occupied by the system up to time t is not sufficient to determine the cell number at time $t+1$ (a unit of discrete time t corresponds to the time interval τ). Therefore, for chaotic orbits, knowledge of the coarse-grained past is not sufficient to determine the coarse-grained future. In contrast, this is possible in non-chaotic systems, since the coarse-grained orbit exhibits regularities. Note that no restrictions on the size of the partition have been made. That is, a sequence of finite precision measurements is unable to predict the future of a chaotic system, independently of their (finite) precision.

Let us illustrate the concept of deterministic chaos by means of an example. We consider the *logistic map*, one of the best known models for studying the transition to chaos. It is defined by the first-order difference equation

$$x_{n+1} = \alpha x_n (1 - x_n),\qquad(1.27)$$

where $0 \le \alpha \le 4$, so that the unit interval $[0, 1]$ is mapped into itself. The behaviour of the logistic map is very complicated and exhibits regions of regular or chaotic motion when the parameter α is varied. In particular, the map is fully chaotic for $\alpha = 4$. Now, let $x_n = \sin^2(\pi y_n)$ and substitute into (1.27) for $\alpha = 4$. After some straightforward algebra we find $\sin^2(\pi y_{n+1}) = \sin^2(2\pi y_n)$. Hence, the logistic map, for $\alpha = 4$, is equivalent to

$$y_{n+1} = 2y_n \quad (\bmod\, 1).\qquad(1.28)$$

This equation maps the unit interval $[0, 1]$ into itself. It has the following simple analytic solution:

$$y_n = 2^n y_0 \quad (\bmod\, 1).\qquad(1.29)$$

A more transparent form of the solution may be obtained by writing the binary representation of y_0:

$$y_0 = 0.1101001100011010\ldots.\qquad(1.30)$$

It is easy to check that each iteration of map (1.28) moves the decimal point, in the binary representation, one digit to the right and then drops the integer part to the left of the decimal point. Therefore, one bit of information is erased at each map step.

It is now easy to show that the solution of the deterministic equation (1.28) is completely unpredictable. In our example the unit interval $[0, 1]$ plays the role of the energy surface (the orbit resides in this interval). We partition this "energy surface" into two cells, a left cell ($0 \leq y < 1/2$) and a right cell ($1/2 \leq y < 1$). From the binary representation (1.30), we recognize that y_n resides in the left or right cell, depending on the value 0 or 1 of the first digit (after the decimal point) of its binary representation. Since the decimal point moves one digit to the right at each map step, the coarse-grained orbit corresponds to the binary representation (1.30) of the initial state y_0: 0 means the left cell and 1 the right cell. It is clear that, if we know the first t numbers of the coarse-grained orbit, we cannot determine the number $t + 1$. Indeed, if we know the first t digits of the binary expansion of y_0, we cannot determine the subsequent digits. As time goes on, the solution will depend on ever diminishing details of the initial condition. In other words, when we fix y_0 we supply the system with infinite complexity which arises owing to the chaotic nature of the motion.

How random is the solution of Eq. (1.28)? Let us assume that someone who knows the precise solution of (1.28) tells us the sequence of digits in y_0. Can we deduce whether this person is really telling us the solution of Eq. (1.28) or merely a sequence of random digits that he has obtained, for instance, by flipping a fair coin? (Say that 0 corresponds to heads, 1 to tails.) The answer is no. Indeed, we can easily convince ourselves that the set of all possible initial conditions y_0 is in one-to-one correspondence with the set of all possible coin-tossing sequences.[1] Since a coin-tossing sequence is random, the binary representation of y_0 is also random. Therefore, the orbit itself is also random.

1.4.2 * *Algorithmic complexity*

At this point we need to clarify exactly what we mean when we say that a digit string is *random*. Each binary digit carries one bit of information. Therefore, an n-bit binary sequence can carry n bits of information. However, if there are correlations between digits, the information contained in this n-bit string can be expressed by a shorter sequence. Following Kolmogorov, we define the *complexity* $K_M(x)$ of a given n-bit sequence x as

---

[1] Note that for chaotic dynamics, the set of coarse-grained orbits is complete, that is, it actually contains all possible sequences.

the bit length of the shortest computer program (algorithm) capable of computing this sequence using machine M. Note that complexity can be made machine independent. Indeed, Kolmogorov has proved the existence of a universal machine U such that

$$K_U(x) \leq K_M(x) + C_M, \qquad (1.31)$$

where C_M depends on M but not on x.

Let us consider the sequence 01010101... This string can be computed by the program "PRINT 01, n times". The length of this program is $\log_2 n + A$, where $\log_2 n$ is the number of bits required to specify n and A is a constant that depends on the machine. Therefore, the complexity of this sequence is $O(\log_2 n)$, independently of the machine used. On the other hand, the complexity of an n-bit string $x = (x_1, x_2, \ldots, x_n)$ can never exceed $O(n)$. Indeed, this sequence can always be produced by the copy program "PRINT $(x_1, x_2, \ldots, x_n)$", which is $O(n)$ bits long. Following Kolmogorov, we say that an n-bit sequence is random if it cannot be calculated by a computer program whose length is smaller than $O(n)$ bits. A chaotic orbit is random in the sense that it cannot be compressed into a shorter sequence; it is therefore unpredictable.

For infinite sequences, we can define the complexity as

$$K_\infty = \lim_{n \to \infty} [K^{(n)}/n], \qquad (1.32)$$

where $K^{(n)}$ is the complexity of the first n bits of the sequence. Note that Kolmogorov's result on the existence of a universal machine tells us that K_∞ is machine independent, this follows trivially from Eq. (1.31). It is possible to show that, in general, limit (1.32) exists. Martin-Löf proved that almost all sequences having positive complexity ($K_\infty > 0$) would pass all computable tests for randomness. This justifies the statement that positive complexity sequences are random. Moreover, Martin-Löf proved that almost all sequences have positive complexity and are therefore random. It follows that almost all orbits that are solutions of map (1.28) are random; their information content is thus both infinite and incompressible.

A further consequence of algorithmic complexity theory is that almost all real numbers cannot be computed by finite algorithms. Of course, exceptions are the integer and rational numbers. We also note that irrational numbers such as π or e are not random, since efficient algorithms to compute them to any desired accuracy exist. These algorithms imply $K_\infty = 0$.

We can now clarify the connection between chaotic dynamics and positive algorithmic complexity. Chaos is defined in terms of sensitivity to initial conditions. If δx_0 is an infinitesimal change of the initial condition for a given dynamical system and δx_t is the corresponding change at time $t > 0$, then in general

$$|\delta x_t| \approx e^{\lambda t}|\delta x_0|, \qquad (1.33)$$

where λ is the largest so-called Lyapunov exponent. We say that the dynamics is chaotic when $\lambda > 0$, that is, when any amount of error in determining the initial conditions diverges exponentially, with rate λ. It is clear from the solution (1.29) of map (1.28) that

$$|\delta y_t| = 2^t|\delta y_0| = e^{(\ln 2)t}|\delta y_0|, \qquad (1.34)$$

and therefore $\lambda = \ln 2$. The exponential sensitivity to initial conditions means that one digit of orbital accuracy is lost per suitably chosen unit of time. To recover this digit of accuracy, we must add one digit of accuracy to the initial condition y_0. Therefore, to be able to follow our orbit up to time t accurately, we must input $O(t)$ bits of information. Thus, a chaotic orbit has positive complexity, namely, it is random.

For non-chaotic systems (in particular, for integrable systems) errors only grow linearly with time and therefore knowledge of the coarse-grained past is sufficient to predict the future.

In conclusion, the solutions of chaotic systems cannot be computed efficiently, since the computational resources required to determine the orbit up to time t grows like the time t itself. Only for non chaotic systems is it possible, at least in principle, to compute the solution efficiently. That is, with computational resources that grow as the input size $\log t$.

1.5 Energy and information

1.5.1 *Maxwell's demon*

In this section, we discuss the connection between energy and information, two concepts that, at first sight, might seem hardly related. We may say that the discussion on the relation between energy and information goes back to *Maxwell's demon* paradox, introduced by James Clerk Maxwell in 1867. He imagined that a demon was capable of monitoring the positions and velocities of the individual molecules of a gas contained in a box and initially in thermal equilibrium at temperature T (see Fig. 1.17). At equi-

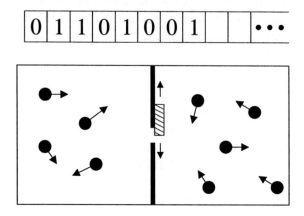

Fig. 1.17 Schematic drawing of Maxwell's demon paradox. The demon's memory is represented by a binary string and stores the results of his measurements of the positions and velocities of the molecules.

librium, the individual molecular velocities are distributed according to the Maxwell distribution and with random directions. The box is divided in two halves, which communicate through a little door. The demon opens and closes the door so as to allow the faster molecules to move from the right half of the box to the left half and the slower molecules to move in the opposite way. By doing this many times the demon separates the faster molecules, which end up on the left side of the box, from the slower molecules, which finish on the right side. As a consequence, the temperature T_l of the gas in the left chamber becomes higher than the temperature T_r of the gas in the right chamber. Since we have obtained two gases at different temperatures, we can now use these two "thermal baths" to produce work. Therefore, the demon has been able to convert heat from a source at uniform temperature T into work, in apparent violation of the second law of thermodynamics. Actually, as we shall see later, there is no such violation, since the transformation of heat into work is not the only result of the process considered as a whole.

Maxwell's demon paradox may be equivalently stated in terms of *entropy*. Entropy is defined as

$$S = k_B \ln \Omega, \qquad (1.35)$$

where $k_B \approx 1.38 \times 10^{-23}$ Joule/K is Boltzmann's constant and Ω is the number of microscopic states of the system (that is, positions and velocities of the single molecules) that give the same macroscopic state (identified by a few parameters, such as the volume V and temperature T). It is clear that the demon introduces order into the system: the faster molecules are constrained to stay on the left side of the container, the slower molecules on the right. Therefore, the number of microscopic states accessible to the system diminishes and the total entropy of the gas is reduced, thus apparently violating the second law of thermodynamics. Moreover, the demon could divide the box into many cells and separate with great accuracy the molecules according to their velocity. As a consequence, the entropy of the gas would become smaller and smaller as the number of cells increased. Indeed, our knowledge of the microscopic state of the system would increase, implying a decrease in the entropy. However, such violation of the second law is only apparent: indeed, a careful analysis shows that the total entropy (of the gas, demon and environment) does not decrease.

1.5.2 *Landauer's principle*

Maxwell's demon spawned much discussion and different solutions were proposed to solve the paradox. At the beginning, it was widely believed that the resolution of the paradox lay in the energy cost of the *measurements* performed by the demon. For example, in order to locate the molecules the demon needs to illuminate them and this has an energy cost. However, Rolf Landauer and Charles Bennett were able to show that the measurement process can, in principle, be performed without energy expenditure. They eventually succeeded in finding the solution of the paradox: the results of the measurements must be stored in the demon's memory. Since his memory is finite, the demon will eventually need to *erase* his memory to free up memory cells for new measurements. There is energy dissipation associated with this erasure. This is the content of Landauer's principle, stated in 1961.

Landauer's principle: *Each time a single bit of information is erased, the amount of energy dissipated into the environment is at least $k_B T \ln 2$, where k_B is Boltzmann's constant and T the temperature of the surrounding environment. Equivalently, we may say that the entropy of the environment increases by at least $k_B \ln 2$.*

Thus, the decrease in the entropy of the gas is compensated by an increase in the demon's entropy. To erase the information gathered by the demon in the measurement process, we must dissipate energy into the environment. Therefore, the energy cost, according to Landauer's principle, is not due to the measurement process itself, but to information erasure.

The following example is useful to illustrate Landauer's principle. Suppose that information is embodied in the state of a physical system, for instance, we might store a bit of information via a single molecule in a box. We say that the bit is set to 0 if the molecule is on the left side of the box, to 1 if it is on the right. Even though we have a single-molecule system, we may apply the laws of thermodynamics. As is well known,

$$dE = \delta L + \delta Q, \tag{1.36}$$

where dE is the variation of the internal energy of the gas, δL the work done on the gas and δQ the heat absorbed by the gas. If we consider a quasi-static transformation (that is, a transformation so slow that we may consider the system to be always in an equilibrium state), we may write

$$dS = \frac{\delta Q}{T}, \tag{1.37}$$

where dS is the variation of the entropy of the gas. Let us assume that our box is in contact with a thermal bath at temperature T and that we compress the gas by means of a frictionless piston (see Fig. 1.18). If the displacement of the piston is dx, the work done on the gas is given by

$$\delta L = -Fdx = -pAdx = -pdV, \tag{1.38}$$

where F is the force of the gas on the piston, p its pressure, A the surface of the piston and V the volume of the gas. Of course, since we only have a single molecule, concepts such as pressure and force must be understood in a time-averaged sense, that is, we need to average over many collisions of the molecule against the piston. Let us consider a transformation that halves the volume of the gas. Taking into account the equation of state for ideal gases,

$$pV = Nk_BT, \tag{1.39}$$

where N is the number of particles in the gas (here $N = 1$), we can compute the work done on the gas as follows:

$$L = -\int_V^{V/2} pdV' = -\int_V^{V/2} \frac{k_BT}{V'} dV' = k_BT \ln 2. \tag{1.40}$$

Note that here we have used the fact that the transformation is isothermal since the system is in contact with a thermal bath at temperature T.

As we have assumed that the gas is ideal, its internal energy does not change because the temperature is constant. Therefore, from the first law of thermodynamics, Eq. (1.36), the work done on the gas is transformed into heat dissipated into the environment: $\Delta Q = -L$. Note that $\Delta Q < 0$ since heat is dissipated, not absorbed. The change in the entropy of the gas is given by Eq. (1.37):

$$\Delta S = \frac{\Delta Q}{T} = \frac{-L}{T} = -k_B \ln 2. \qquad (1.41)$$

We have $\Delta S < 0$ since, after compression, the volume available to the molecule is halved and therefore the number of available microstates is reduced correspondingly. The entropy of the system diminishes, while the entropy of the environment ΔS_{env} increases: since the total entropy of the universe can never decrease, we have $\Delta S + \Delta S_{\text{env}} \geq 0$. Thus, $\Delta S_{\text{env}} \geq k_B \ln 2$, in agreement with Landauer's principle.

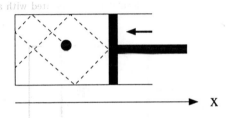

Fig. 1.18   Compression of a single-molecule gas by means of a piston.

Let us now assume that a binary message is stored by means of a sequence of single-molecule boxes. Each box carries a single bit of information, set in the state 0 or 1, depending on the left/right position of the molecule (see Fig. 1.19). We shall now show the validity of the following statement: the information contained in the message is proportional to the energy necessary to erase the message, that is, to move all the molecules to the left (or right) side of the boxes. First of all, we must define the information content of a message. It is defined as the information we should gain if we knew the values of the bits that constitute the message. Therefore, information is a measure of our *ignorance* about the message. If we already knew the values of the bits, we should obtain no further information from the message. In this case the information contained in the message is zero and, according to the statement made above, no energy expenditure is nec-

essary to erase the message. Let us show that indeed no work is required
to set the state of each bit to 0. Indeed, if the molecule is already on the
left side, no further action is required. On the other hand, if the molecule
is on the right side, we can move it to the left without energy expenditure:
it is sufficient to enclose it inside a smaller, inner box and to shift this box
to the left, as shown in Fig. 1.20. No work is required to perform this
operation, since the molecule bounces as many times against the left wall
of the inner box as against the right. Only in the case in which we do not
know in advance the position of the molecule must we halve the volume of
the gas and, as we have seen previously, this requires work $L = k_B T \ln 2$
to be done on the gas. In this latter case the information content of the
message is different from zero and we must expend energy to erase it.

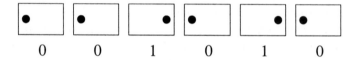

$$0 \qquad 0 \qquad 1 \qquad 0 \qquad 1 \qquad 0$$

Fig. 1.19   A sequence of single-molecule boxes associated with a binary string.

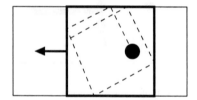

Fig. 1.20   A procedure enabling the transfer of a molecule to the left side of a box
without energy cost.

1.5.3   *Extracting work from information*

For a better understanding of the relation between information and energy,
it is instructive to consider the following example, devised by Bennett,
which shows that information may be used as fuel to move a machine. A
trolley is in contact with a thermal bath at temperature T and a ribbon,
made of a string of single-molecule boxes, enters the trolley (see Fig. 1.21).
If we know in advance the left/right position of every molecule, we can
extract work to move the trolley. Indeed, it is sufficient to insert a piston
in the middle of each box. As shown in Fig. 1.22, the piston is movable to

the right if the molecule is on the left side of the box and to the left if the molecule is on the right side. Since the whole system is at temperature T, we extract work $L = k_B T \ln 2$. If we have an N-bit ribbon, the total work is $N k_B T \ln 2$ and it may be used to displace the trolley. We stress that, when the ribbon comes out of the trolley, the molecules can be anywhere inside the volume V. The information content of the string of boxes has been completely lost and used as a fuel to move the trolley. On the other hand, if we do not know in advance the left/right position of the molecules, then we cannot extract useful work: indeed, if we insert a piston, half of the time the gas produces work, but the other half of the time work is done on the gas. On average, the extracted work is thus zero.

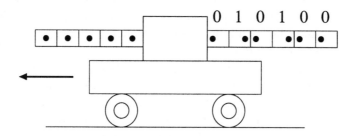

Fig. 1.21   Use of information to produce work.

Fig. 1.22   Extraction of work from a single-molecule gas, which at the beginning is on the left or right side of a container.

1.6   Reversible computation

In this section we discuss the energy requirements for computation. Most of the logic gates introduced in Sec. 1.2 are *irreversible*. This means that, given the output, we cannot recover the input. For instance, if the OR gate has output 1, the input bits could have been set to $(0,1)$, $(1,0)$ or $(1,1)$. The Boolean functions $f : \{0,1\}^2 \to \{0,1\}$ erase a bit of information and therefore, according to Landauer's principle, the amount of energy dissipated into the environment must be at least $k_B T \ln 2$. This is analogous to the example of a molecule in a box discussed in the previous section: instead of halving the volume accessible to a gas, we pass from a two-bit input to a single-bit output.

We note that the value $k_B T \ln 2$ represents only a lower bound for the energy dissipation of a two-bit irreversible gate. Present-day, real computers dissipate more by orders of magnitude. However, the energy dissipation per gate has been reduced enormously over the years thanks to technological progress. On the other hand, if we increase the power of a computer (that is, the number of operations per second), we also increase the energy dissipated, unless we are able to reduce the energy cost of each elementary logic gate. It is therefore important to keep in mind that Landauer's principle sets a lower bound to any future reduction of the energy dissipated by irreversible computation.

Since the energy cost of computation is related to irreversibility, the following question arises: is it possible to build a reversible computation without energy consumption? We can see in advance that it should be possible, since the fundamental laws of physics (Newton's equations in classical mechanics) are reversible. Therefore, there must be some underlying reversible physical process that allows us to implement irreversible gates. Indeed, any irreversible function $f : \{0,1\}^m \to \{0,1\}^n$ can be embedded into a reversible function. It is sufficient to define the function

$$\tilde{f} : \{0,1\}^{m+n} \to \{0,1\}^{m+n}, \tag{1.42}$$

such that

$$\tilde{f}(x,y) = (x, [y + f(x)] \,(\mathrm{mod}\, 2^n)), \tag{1.43}$$

where x represents m bits, while y and $f(x)$ represent n bits. Since $\tilde{f}$ takes distinct inputs into distinct outputs, it is an invertible $(m+n)$-bit function.

According to the above argument, it is possible to find universal reversible gates for computation. We shall indeed show that the universal gates NAND and FANOUT can be constructed from reversible gates. In order to avoid information loss, a reversible function must take n input bits into n output bits. Reversible functions are permutations of the 2^n possible inputs and therefore their number is $(2^n)!$. For $n = 1$ there are 2 reversible single-bit gates: the identity and the NOT gate. A very important two-bit gate is the controlled-NOT (CNOT) or reversible XOR, shown in Fig. 1.23.

The first bit (a) acts as a control and its value is unchanged on output: $a' = a$. The second (target) bit is flipped if and only if the first bit is set to one and therefore $b' = a \oplus b$. Therefore, on output the second bit provides the XOR of the inputs a and b. Furthermore, the CNOT gate is reversible since from the output (a', b') we may infer the input (a, b). Note that, if we set the target bit to 0, the CNOT gates becomes the FANOUT gate: $(a, 0) \to (a, a)$. It is easy to check that CNOT is self-inverse. Indeed, the application of two CNOT gates, one after the other, leads to

$$(a, b) \to (a, a \oplus b) \to (a, a \oplus (a \oplus b)) = (a, b). \tag{1.44}$$

Therefore, $(\text{CNOT})^2 = I$, that is, $\text{CNOT}^{-1} = \text{CNOT}$.

An interesting consequence of the fact that the FANOUT gate can be obtained from the CNOT is that a measurement process is, in principle, possible without energy expenditure. Indeed, a measurement establishes a correlation between the state of a system and the state of a memory register. Therefore, it is equivalent to a copying operation (FANOUT), which can be performed reversibly. This argument shows that the solution of Maxwell's demon paradox does not indeed lie in the measurements performed by the demon.

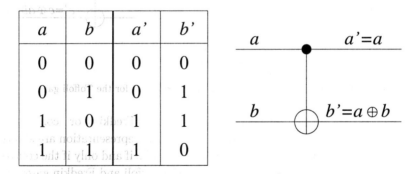

a	b	a'	b'
0	0	0	0
0	1	0	1
1	0	1	1
1	1	1	0

Fig. 1.23   The truth table and circuit representation for the CNOT gate.

1.6.1 *Toffoli and Fredkin gates*

It is possible to show that two-bit reversible gates are not enough for universal computation (see, *e.g.*, Preskill, 1998). Instead, a universal gate is the controlled-controlled-NOT (CCNOT) or Toffoli gate, which is a three-bit gate. Its truth table and circuit representation are shown in Fig. 1.24. This gate acts as follows: the two control bits are unchanged ($a' = a$ and $b' = b$) while the target bit is flipped if and only if the two control bits are set to 1, that is, $c' = c \oplus ab$. In order to prove that the Toffoli gate is universal, we shall use it to construct both NAND and FANOUT gates. Indeed, if we set $a = 1$, the Toffoli gate acts on the other two bits as a CNOT and we have seen that the FANOUT gate can be constructed from the CNOT. To construct the NAND gate, we set $c = 1$, so that $c' = 0$ if and only if $a = 1$ and $b = 1$, that is, $c' = 1 \oplus ab = a$ NAND b

Exercise 1.2 Construct the NOT, AND, OR gates from the Toffoli gate.

a	b	c	a'	b'	c'
0	0	0	0	0	0
0	0	1	0	0	1
0	1	0	0	1	0
0	1	1	0	1	1
1	0	0	1	0	0
1	0	1	1	0	1
1	1	0	1	1	1
1	1	1	1	1	0

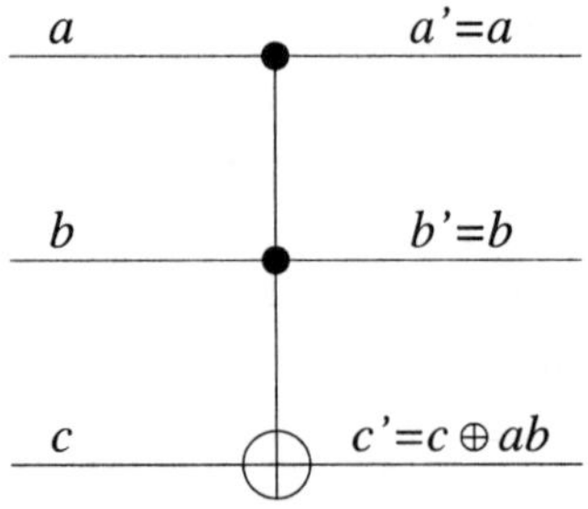

Fig. 1.24   The truth table and circuit representation for the Toffoli gate.

Another universal reversible gate is the Fredkin or controlled-EXCHANGE gate, whose truth table and circuit representation are shown in Fig. 1.25. This gate swaps the input bits b and c if and only if the control bit a is set to 1. As is easy to check, both the Toffoli and Fredkin gates are self-inverse.

Exercise 1.3 Show that the Fredkin gate is universal.

We have seen that irreversible gates, such as AND and OR, can be embedded into reversible gates. However, the price to pay is the introduction of additional bits and on output this produces "garbage" bits, which are not reused during the computation. These extra bits are needed to store the information that would allow us to reverse the operations. For instance, if we set $c = 1$ at the input of the Toffoli gate, we obtain $c' = a$ NAND b

a	b	c	a'	b'	c'
0	0	0	0	0	0
0	0	1	0	0	1
0	1	0	0	1	0
0	1	1	0	1	1
1	0	0	1	0	0
1	0	1	1	1	0
1	1	0	1	0	1
1	1	1	1	1	1

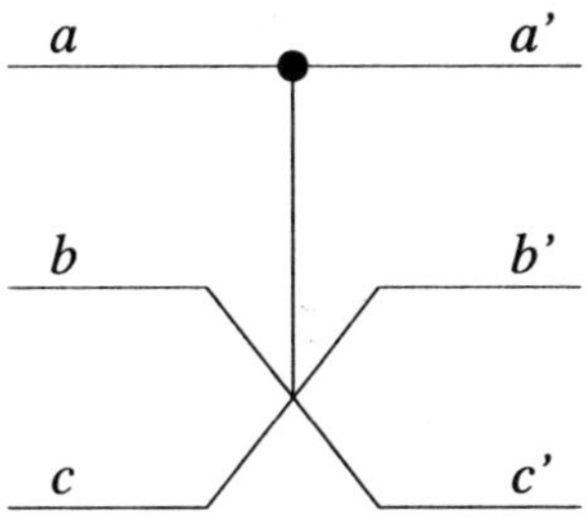

Fig. 1.25   The truth table and circuit representation for the Fredkin gate.

plus two garbage bits ($a' = a$ and $b' = b$). One might think that energy is required to erase this garbage, thus nullifying the advantage of reversible computation. Fortunately, as was shown by Bennett, this is not the case. Indeed, we can perform the required computation, print the result and then run the computation backward, again using reversible gates, to recover the initial state of the computer. As a consequence, the garbage bits return to their original state without any energy consumption.

1.6.2   *⋆ The billiard-ball computer*

A concrete example of reversible computation is the billiard-ball computer. In this computer the value taken by a bit is associated with the absence (0) or the presence (1) of a ball in a given position. The transmission of information is performed by means of frictionless motion of the balls on a plane surface (a billiard table) and the logic gates are implemented by means of elastic collisions between balls and against fixed obstacles. The positions of the balls on the left- and right-hand sides of the billiard table give the input and output, respectively. As an example, let us consider the collision gate depicted in Fig. 1.26. In this figure, on input we have a ball in a and another in b, that is, $a = b = 1$. After the collision, the balls are recovered in a' and b'; that is, $a' = b' = 1$, while $a'' = b'' = 0$. If instead we have zero or a single ball, then there are no collisions. For instance, if $a = 1$ and $b = 0$, then $a'' = 1$ and $a' = b' = b'' = 0$. Therefore, the collision gate computes the following logical functions: $a' = b' = a \wedge b$, $a'' = a \wedge \bar{b}$

and $b'' = \bar{a} \wedge b$. To implement a universal reversible computation, we also need the possibility to change the direction of the balls. This is obtained by means of elastic bounces off a fixed obstacle (see Fig. 1.27).

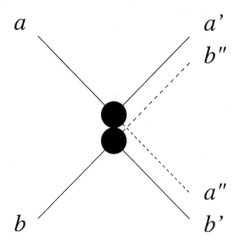

Fig. 1.26   A collision gate in a billiard-ball computer.

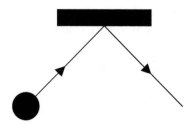

Fig. 1.27   Change of direction of a ball in a billiard-ball computer.

It is possible to combine the basic elements depicted in Figs. 1.26 and 1.27 to implement the Fredkin gate (see, for example, Feynman, 1996). Since the Fredkin gate is universal, it turns out that the billiard-ball computer may be used to compute any complex Boolean function. However, it is also clear that the actual implementation of such a computer is hindered by the instability of the motion of the balls under perturbations. Finally, it is interesting to note that the billiard-ball computer, besides being reversible, is also conservative. This means that the number of balls on input is equal to the number of balls on output; that is, the number of 0's and

1's is conserved. Actually, this is one of the properties of the Fredkin gate.

1.7 A guide to the bibliography

The Turing machine was introduced in Turing (1936) and the Church-Turing thesis first stated in Church (1936).

Classical books on algorithm design are Cormen *et al.* (2001) and Knuth (1997–98).

Textbooks on computational complexity are Garey and Johnson (1979) and Papadimitriou (1994). An informal introduction to the theory of computational complexity, directed at physicists, is Mertens (2000). The computational complexity of quantum computing is discussed in Bernstein and Vazirani (1997).

The algorithmic complexity of dynamical systems is reviewed in Alekseev and Jacobson (1981) and a very readable discussion can be found in Ford (1983).

A profound discussion of the relation between energy and information is given by Feynman (1996). Landauer's principle was stated in Landauer (1961). Maxwell's demon paradox has been reviewed by Bennett (1982) and Bennett (1987).

Reversible computation is discussed in Bennett (1973) and Fredkin and Toffoli (1982).

itself, however. Actually, this is one of the properties of the Lagrangian.

1.7 A guide to the bibliography

Chapter 2

Introduction to Quantum Mechanics

At the end of nineteenth century it became clear that classical physics led to predictions in disagreement with experiment. This gave rise to a profound change in the basic concepts of our understanding of Nature. A new theory, known as quantum mechanics, was constructed. This theory describes the phenomena of the microscopic world in satisfactory agreement with all present experimental data.

This chapter is a simple introduction to quantum mechanics. Our aim is to provide the necessary background for an understanding of the subsequent chapters. Note that no prior knowledge of quantum mechanics is required. Moreover, it is easy to learn the basic aspects of quantum mechanics. Certainly, it is much easier than one might think, considering the development of sophisticated quantum-mechanical techniques for the understanding of complex phenomena or the counter-intuitive, even paradoxical, consequences of quantum mechanics.

We begin with the description of two simple yet classic experiments: the Stern–Gerlach experiment and Young's double-slit experiment, illustrating the distinctive features of quantum mechanics. Then we review the basics of elementary linear algebra, since its main concepts are necessary for an understanding of quantum mechanics. For our purposes, it will be sufficient to consider finite-dimensional vector spaces. After that, we describe the postulates of quantum mechanics. We shall confine ourselves to the case of systems described by wave vectors residing in finite-dimensional Hilbert spaces. Finally, we elucidate the unusual, non-classical properties of quantum mechanics. We discuss the EPR paradox and Bell's inequalities, a spectacular example of the profound difference between quantum and classical physics.

2.1   The Stern–Gerlach experiment

In this section, we give a simple description of the Stern–Gerlach experiment. Perhaps, this is the experiment that illustrates most dramatically the inadequacy of classical mechanics to describe physical phenomena. It forces us to think in terms of quantum mechanics and to give up the traditional classical description of Nature. Indeed, the Stern–Gerlach experiment exhibits a typical quantum-mechanical behaviour and the predictions of classical mechanics are invalidated. It therefore shows that the basic concepts of classical mechanics must be modified in order to understand certain physical phenomena.

The Stern–Gerlach apparatus is shown schematically in Fig. 2.1: a beam of neutral atoms, having magnetic moment μ, enters a region in which there is a magnetic field B directed along the z-axis. The magnetic field is inhomogeneous, with the gradient ∇B directed along z (B denotes the modulus of the vector B). Under these conditions, classical mechanics tells us that the atoms are subjected to a force F, also directed along z. If F_z and μ_z denote the projections of F and μ along z, we have

$$F_z = \mu_z |\nabla B| = \mu_z \frac{dB}{dz}. \tag{2.1}$$

The atoms are deflected with respect to the incoming direction by the gradient of the magnetic field and then reach a screen S. If we measure the deflection on the screen, we can derive the force F_z and therefore the component μ_z of the magnetic moment of the atoms.

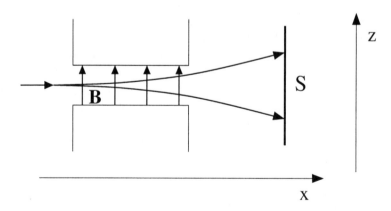

Fig. 2.1   Schematic drawing of the Stern–Gerlach experiment.

At the entrance to the region with the magnetic-field gradient, the magnetic moments of the atoms are distributed isotropically. Therefore, according to classical mechanics, all values of μ_z between $-m$ and $+m$, with $m \equiv |\mu|$, are allowed. As a consequence, the impact points of the atoms on the screen should be distributed continuously around the incoming direction, with maximum positive and negative deflections corresponding to the values $\mu_z = +m$ and $\mu_z = -m$. However, the experimental results are in clear contradiction with such predictions. Only a finite number of spots are registered on the screen. These spots are equally spaced along z and contained within the interval between the maximum positive and negative deflections corresponding to $\mu_z = m$ and $\mu_z = -m$, respectively. This means that the allowed values of μ_z are discrete. In some cases, for instance when using silver atoms, there are only two spots on the screen, corresponding to $\mu_z = -m$ and $\mu_z = m$.

This apparently mysterious phenomenon found its explanation in the fact that electrons possess intrinsic angular momentum, known as *spin*. The magnetic moment of the atom is proportional to its angular momentum and, for atoms like silver, the angular momentum is simply equal to the spin of the outer electron. The two spots on the screen correspond to the two allowed spin states with respect to a given direction, which may be labelled *spin up* and *spin down*. The experimentally determined values of the spin angular momentum are given by $S_z = +\frac{1}{2}\hbar$ (if the spin is up) and $S_z = -\frac{1}{2}\hbar$ (if the spin is down), where $h = 2\pi\hbar$ is Planck's universal constant, whose value is given by $h \approx 6.626 \times 10^{-34}$ Joule sec. Therefore, we say that the electron is a spin-$\frac{1}{2}$ particle. Note that the z direction in the Stern–Gerlach experiment is arbitrary; the same results are obtained if the magnetic field is oriented along any direction.

We now consider the experiment drawn schematically in Fig. 2.2. The first apparatus splits the initial beam into two components, corresponding to the spin-up and spin-down states of the electrons. Using a notation whose meaning will become clear later in this chapter, we call these two components $|+\rangle_z$ and $|-\rangle_z$, Then we block the component $|-\rangle_z$ and let the other component $|+\rangle_z$ enter a second Stern–Gerlach apparatus analogous to the first one, namely, with the magnetic field oriented along the z axis. A single beam, corresponding to the component $|+\rangle_z$, is observed to come out of the second apparatus. The $|-\rangle_z$ component is not present, since it has been previously cut off. In short, the first Stern–Gerlach apparatus filters the atoms and only those with $\mu_z = m$ are selected. Thus, a second Stern–Gerlach apparatus measures this same spin component.

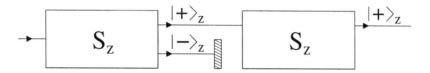

Fig. 2.2   Sketch of a Stern–Gerlach experiment. The first apparatus filters out the atoms with $\mu_z = -m$ while the second measures μ_z, obtaining $\mu_z = m$.

A different arrangement of the Stern–Gerlach experiment is illustrated in Fig. 2.3. Unlike the case of Fig. 2.2, the magnetic field in the second apparatus is directed along the y axis. Now we observe that two beams with equal intensity emerge from the second apparatus, corresponding to $\mu_y = +m$ and $\mu_y = -m$, where μ_y denotes the projection of the magnetic moment of the atoms along y. We call these two components $|+\rangle_y$ and $|-\rangle_y$. This result is not surprising since the atoms enter the second Stern–Gerlach apparatus with a well-determined $\mu_z = m$, but the value of μ_y is not given. Is it therefore correct to say that half of the atoms that enter the second apparatus have components $|+\rangle_z$ and $|+\rangle_y$ and the other half have components $|+\rangle_z$ and $|-\rangle_y$? As illustrated by the following experiment, this intuitive way of thinking is *not* valid.

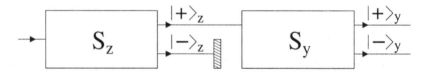

Fig. 2.3   Sketch of a Stern–Gerlach experiment. The first apparatus filters out atoms with $\mu_z = -m$ while the second measures μ_y.

Indeed, a very surprising result is obtained using the experimental setup drawn in Fig. 2.4. In this case, the first two Stern–Gerlach apparatuses filter out the atoms with magnetic-moment components $\mu_z = -m$ and $\mu_y = -m$. The amazing result is that both components $|+\rangle_z$ and $|-\rangle_z$ come out with equal intensity from the third apparatus, even though the component $|-\rangle_z$ was previously filtered out. How then is it possible that the component $|-\rangle_z$ reappears after the third apparatus? Where does this component come from? This experiment shows that it is not correct to think that the atoms entering the third apparatus are in the state $|+\rangle_z$ and $|+\rangle_y$. It is also extremely puzzling that, if the plate screening the $|-\rangle_y$ component is removed, so that both components $|+\rangle_y$ and $|-\rangle_y$ enter the third apparatus, then only the component $|+\rangle_z$ comes out. The experiment drawn in Fig. 2.4 is an impressive illustration of a fundamental property of quantum mechanics: the final state of the system depends only on the state of the atoms that enter the last Stern–Gerlach apparatus and on the action of this apparatus; there is no memory of the previous history of the system. In short, the second apparatus singles out the state $|+\rangle_y$ and in so doing completely destroys any information about the value of S_z. In the case of Fig. 2.4 the atoms that enter the last apparatus satisfy $\mu_y = m$, but there are no restrictions on μ_z. Therefore, some of the atoms come out of the apparatus with $\mu_z = +m$, others with $\mu_z = -m$.

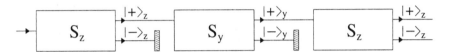

Fig. 2.4   Sketch of a Stern–Gerlach experiment. The first apparatus filters out the atoms with $\mu_z = -m$, the second those with $\mu_y = -m$, the third measures μ_z.

2.2   Young's double-slit experiment

Another experiment which effectively illustrates the distinctive features of quantum mechanics is Young's double-slit experiment. As is well known, the nature of light has been the focus of deep debate over the centuries. The question was if light is a beam of particles or a wave. Newton supported the particle conception of light, since he believed that the presence of a sharp shadow behind an object could not be explained if light were a wave. Therefore, he concluded that light is not a wave like sound, which can be heard even behind objects. However, during the nineteenth century, the wave-like nature of light was demonstrated in interference experiments. Young formulated the *superposition principle*: if two waves, emitted by a single source, fall on a screen, their *amplitudes* (not their intensities, which are square moduli of the amplitudes) add up algebraically. This property is at the origin of the well-known interference fringes observed in the double-slit experiment illustrated in Fig. 2.5. Light is emitted by the source S, passes through two slits, O_1 and O_2, and illuminates a screen (a photographic plate, for example). Typical interference fringes are produced on the screen, as sketched in Fig. 2.5. The main point is that the intensity $I(x)$ of light on the screen is different from the algebraic sum of the intensities $I_1(x)$ and $I_2(x)$ produced by the two slits separately, that is, when we close slit O_2 or O_1, respectively. Thus, we have

$$I(x) \neq I_1(x) + I_2(x).\qquad(2.2)$$

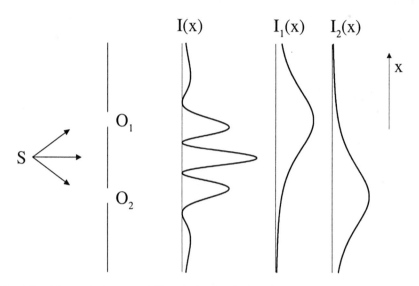

Fig. 2.5   Schematic diagram of Young's double-slit interference experiment. A source S emits light, which can pass through two slits O_1 and O_2, before striking a screen. The pattern $I_1(x)$ is produced when only slit O_1 is open and pattern $I_2(x)$ is obtained when only O_2 is open. The intensity $I(x)$ of light on the screen when both slits are open is different from the algebraic sum of the intensities $I_1(x)$ and $I_2(x)$.

In the second half of the nineteenth century, after the synthesis performed by Maxwell, it became clear that light is an electromagnetic wave. In this framework, the speed of light ($c \approx 2.998 \times 10^8$ m/sec in vacuum) is the propagation velocity of the electromagnetic field and is related to certain electric and magnetic constants. However, the energy distribution of the radiation emitted by a black body could not be explained by the electromagnetic theory. This problem led Planck in 1900 to introduce the hypothesis that light is emitted or absorbed only in integer multiples of a basic *quantum of energy*

$$E = h\nu,\qquad(2.3)$$

where ν is the frequency of light and h is Planck's constant. In 1905 Einstein, in his theory of the photoelectric effect, returned to the particle theory: light consists of a beam of particles, called *photons*, each possessing energy $h\nu$ and momentum $p = h\nu/c$. Both Maxwell's theory, which describes light as an electromagnetic wave and Einstein's theory, which describes light as a beam of elementary particles called photons, received broad experimental confirmation. This again raised the question as to the particle or wave-like nature of light.

The crucial relevance of Young's double-slit experiment lies in the fact that a complete description can be obtained only by accepting simultaneously *both* the wave and particle aspects of light. The light intensity $I(x)$ at a point x on the screen is proportional to the squared modulus of the electric field $E(x)$ at the same point. Let us denote by $E_1(x)$ and $E_2(x)$ the electric field produced at x by beams passing through slits O_1 and O_2, respectively. The corresponding light intensities are given by

$$I_1(x) \propto \left| E_1(x) \right|^2 , \qquad (2.4a)$$

which is observed when the second slit is closed, and

$$I_2(x) \propto \left| E_2(x) \right|^2 , \qquad (2.4b)$$

obtained when the first slit is closed. If we open both slits, the field strengths add up algebraically,

$$E(x) = E_1(x) + E_2(x) , \qquad (2.5)$$

and therefore the resulting light intensity is given by

$$I(x) \propto \left| E(x) \right|^2 = \left| E_1(x) + E_2(x) \right|^2 , \qquad (2.6)$$

and thus we find

$$I(x) \neq I_1(x) + I_2(x) . \qquad (2.7)$$

This is in agreement with the predictions of the wave theory of light.

What happens though if the light intensity is reduced, so that the source only emits photons one-by-one? In this case, each photon produces a *localized* impact at some point on the screen. If we expose the photographic plate for a time so short that only a few photons strike the screen, we observe a few localized impact points, but not an interference pattern. Therefore, a particle interpretation rather than a wave interpretation of light explains this experimental result.[1] Indeed, if we consider a wave, when its intensity

---

[1]Note, however, that the arrival points are not as predicted by classical mechanics, but are distributed probabilistically according to the fringe-pattern intensity.

diminishes, the interference fringes diminish in intensity but do not disappear. Thus, this prediction of the wave theory of light is invalidated by experimental results.

Nevertheless, if the exposure time is sufficiently long that the photographic plate can capture many photons, the interference fringes do appear. The light intensity collected at any given point is proportional to the density of photon impacts at this point. Therefore, the photons, as they arrive, build up the interference fringes and these fringes can be explained by a wave interpretation instead of a particle interpretation of light. In summary, we cannot explain the whole of the experimental results using either the predictions of the particle theory alone or those of the wave theory alone. Both the particle and wave-like nature of light are present.

It is important to note that, if we place a photon counter behind one of the two slits, we indeed determine through which slit each photon passes, but in so doing we destroy the interference fringes. The crucial point is that no experiment can both observe the interference pattern and determine through which slit each photon passes.

These results force us to revise some of the fundamental concepts of classical physics. The fact that interference fringes are observed if and *only* if we do not know through which slit each photon passes forces us to give up the concept of a *trajectory*. Indeed, each photon produced by the source strikes the screen at a different position. We cannot predict this position, but only give the probability $p(x)$ that a photon will strike the screen at the point x. This probability is proportional to the intensity $I(x)$, namely, to $|E(x)|^2$. Even though all photons are emitted under the same conditions, we cannot know in advance where each photon will strike the screen. Therefore, we must give up the classical concept that, once the initial conditions and external forces acting on a particle are known, we can follow, at least in principle, the evolution in time of the particle's coordinates.

We must also give up the concept that the particle and wave descriptions are mutually exclusive, since both aspects are necessary to explain the experimental results. Therefore, we are inevitably led to the concept of *wave–particle duality*: light behaves simultaneously both as a wave and as a particle. We stress that, as has been shown by an overwhelming number of experiments, such duality is not restricted to the description of optical phenomena. Indeed, material particles may also exhibit wave properties and, *vice versa*, waves can be associated with particles. Finally, we would like to emphasize that this radical revision of the concepts of classical physics was imposed and guided by experimental results.

2.3 Linear vector spaces

In this section, we review some elementary linear algebra. We shall neither be concerned with mathematical rigour nor completeness of exposition, but shall just provide the basic notions required to understand the fundamental principles of quantum mechanics. This section is self-contained and no previous knowledge of linear algebra is required. The reader already familiar with the basic elements of linear algebra can therefore pass directly to Sec. 2.4. Nevertheless, a quick glance at the present section may be useful for readers who are not familiar with Dirac's 'bra–ket' notation, which will be adopted throughout this book.

We are interested in *finite-dimensional complex linear vector spaces.* The elements of a vector space V are called *vectors.* An example of a vector space is given by the space C^n in which a vector is singled out by an n-tuple of complex numbers $(\alpha_1, \alpha_2, \ldots, \alpha_n)$, where n is the dimension of the vector space. Following Dirac's notation, we write a vector using the symbol $|\alpha\rangle$ and call it a *ket.*

Two vectors (kets) $|\alpha\rangle, |\beta\rangle \in V$ may be added to give a new vector

$$|\gamma\rangle = |\alpha\rangle + |\beta\rangle, \tag{2.8}$$

residing in the same linear vector space V. In the vector space C^n, we have, in terms of the vector components $|\alpha\rangle = (\alpha_1, \ldots, \alpha_n)$, $|\beta\rangle = (\beta_1, \ldots, \beta_n)$ and $|\gamma\rangle = (\gamma_1, \ldots, \gamma_n)$,

$$\gamma_i = \alpha_i + \beta_i, \qquad (i = 1, \ldots, n). \tag{2.9}$$

Vector addition has the following properties:

$$|\alpha\rangle + |\beta\rangle = |\beta\rangle + |\alpha\rangle, \tag{2.10a}$$

$$|\alpha\rangle + \big(|\beta\rangle + |\gamma\rangle\big) = \big(|\alpha\rangle + |\beta\rangle\big) + |\gamma\rangle. \tag{2.10b}$$

It is possible to multiply a vector $|\alpha\rangle \in V$ by a complex number $c \in C$, to obtain a new vector $c|\alpha\rangle$. The following properties hold for any $c, d \in C$

and $|\alpha\rangle$, $|\beta\rangle \in V$:

$$c\big(|\alpha\rangle + |\beta\rangle\big) = c|\alpha\rangle + c|\beta\rangle\,, \tag{2.11a}$$

$$(c+d)|\alpha\rangle = c|\alpha\rangle + d|\alpha\rangle\,, \tag{2.11b}$$

$$(cd)|\alpha\rangle = c(d|\alpha\rangle)\,. \tag{2.11c}$$

A vector space contains the zero vector 0, which is defined by the following requirement: for any vector $|\alpha\rangle$ belonging to the vector space, $|\alpha\rangle + 0 = |\alpha\rangle$. Note that for the zero vector we do not use the ket notation, since, as we shall see later, $|0\rangle$ means something else, a state of the computational basis. We note that in a vector space $0|\alpha\rangle = 0$, $1|\alpha\rangle = |\alpha\rangle$ and $|\alpha\rangle - |\alpha\rangle = 0$.

Linear independence

A set of vectors $|\alpha_1\rangle, \ldots, |\alpha_m\rangle \in V$ are said to be linearly independent if the relation

$$c_1|\alpha_1\rangle + c_2|\alpha_2\rangle + \cdots + c_m|\alpha_m\rangle = 0\,, \tag{2.12}$$

with $c_1, c_2, \ldots, c_m$ complex numbers, holds *if and only if* $c_1 = c_2 = \cdots = c_m = 0$.

Inner product

The inner product of an ordered pair of vectors $|\alpha\rangle$, $|\beta\rangle \in V$ is a complex number, denoted as $\langle\alpha|\beta\rangle$, with the following requirements:

(i) $\qquad \langle\alpha|\beta\rangle = \langle\beta|\alpha\rangle^\star \qquad$ (skew symmetry), $\qquad$ (2.13a)

where, for any complex number $c = a + ib$ $(a, b \in \mathbf{R})$, $c^\star = a - ib$ denotes its complex conjugate;

(ii) $\quad \langle\alpha|c\beta + d\gamma\rangle = c\langle\alpha|\beta\rangle + d\langle\alpha|\gamma\rangle \quad$ (linearity), $\qquad$ (2.13b)

with $|\alpha\rangle$, $|\beta\rangle$, $|\gamma\rangle \in V$ and $c, d \in \mathbf{C}$;

(iii) $\qquad \langle\alpha|\alpha\rangle \geq 0 \qquad\qquad$ (positivity), $\qquad$ (2.13c)

for any $|\alpha\rangle \in V$, with equality if and only if $|\alpha\rangle$ is the zero vector.

Taking into account the previous relations, it is easy to check the following property:

$$\langle c\alpha|\beta\rangle = c^\star\langle\alpha|\beta\rangle\,. \tag{2.14}$$

Note that $\langle\alpha|$ is the *dual vector* (also called *bra*) to the vector $|\alpha\rangle$; the dual vector $\langle\alpha|$ is a linear operator from the vector space V to the complex numbers C, defined by $\langle\alpha|(|\beta\rangle) = \langle\alpha|\beta\rangle$, for any $|\beta\rangle \in V$.

As an example, we can define an inner product between two vectors $|\alpha\rangle = (\alpha_1, \ldots, \alpha_n)$ and $|\beta\rangle = (\beta_1, \ldots, \beta_n)$ in C^n as follows:

$$\langle\alpha|\beta\rangle = \sum_{i=1}^{n} \alpha_i^{\star}\beta_i. \tag{2.15}$$

The norm

The norm of a vector $|\alpha\rangle$ is defined by

$$\||\alpha\rangle\| = \sqrt{\langle\alpha|\alpha\rangle}. \tag{2.16}$$

It is possible to normalize any non-zero vector $|\alpha\rangle$ by dividing by its norm $\||\alpha\rangle\|$. The normalized vector $|\alpha\rangle/\||\alpha\rangle\|$ has unit norm and is therefore called a *unit vector*. Using the inner product in C^n defined above, then the norm of a vector $|\alpha\rangle = (\alpha_1, \ldots, \alpha_n)$ is given by

$$\||\alpha\rangle\| = \sqrt{\sum_{i=1}^{n} |\alpha_i|^2}, \tag{2.17}$$

and a unit vector must satisfy the condition $\sum_i |\alpha_i|^2 = 1$.

We shall see that quantum mechanics associates with a physical system a unit vector residing in a *Hilbert space*. In the finite-dimensional case, which is the case relevant for quantum information theory as illustrated in this book, a Hilbert space is exactly a complex vector space equipped with an inner product.

The Cauchy–Schwarz inequality

For any two vectors $|\alpha\rangle$ and $|\beta\rangle$,

$$\left|\langle\alpha|\beta\rangle\right|^2 \leq \langle\alpha|\alpha\rangle\langle\beta|\beta\rangle. \tag{2.18}$$

Proof. The inner product is positive definite and therefore $\langle\alpha - c\beta|\alpha - c\beta\rangle \geq 0$ holds for any $|\alpha\rangle, |\beta\rangle \in V$ and $c \in C$. Owing to the linearity of the inner product, this relation is equivalent to $\langle\alpha|\alpha\rangle - c\langle\alpha|\beta\rangle - c^{\star}\langle\beta|\alpha\rangle + cc^{\star}\langle\beta|\beta\rangle \geq 0$. Taking $c = \langle\beta|\alpha\rangle/\langle\beta|\beta\rangle$, we obtain the Cauchy–Schwarz inequality (2.18). $\qquad\square$

Note that, in the special case in which the inner product is real, the Cauchy–Schwarz inequality admits a simple geometrical interpretation. Indeed, since in this case

$$-1 \leq \frac{\langle \alpha | \beta \rangle}{\| |\alpha\rangle \| \, \| |\beta\rangle \|} \leq 1, \tag{2.19}$$

we can write

$$\langle \alpha | \beta \rangle = \| |\alpha\rangle \| \, \| |\beta\rangle \| \cos \theta. \tag{2.20}$$

This latter equation corresponds to the usual definition of the scalar product of two vectors $|\alpha\rangle$ and $|\beta\rangle$, where θ is the angle between the two vectors.

Orthonormality condition

Two non-zero vectors $|\alpha\rangle$ and $|\beta\rangle$ are said to be *orthogonal* if their inner product is zero:

$$\langle \alpha | \beta \rangle = 0. \tag{2.21}$$

A set of vectors $|\alpha_1\rangle, |\alpha_2\rangle, \ldots, |\alpha_n\rangle$ is said to be *orthonormal* if

$$\langle \alpha_i | \alpha_j \rangle = \delta_{ij} \qquad (i, j = 1, 2, \ldots, n), \tag{2.22}$$

where δ_{ij} is the Kronecker symbol, defined as $\delta_{ij} = 1$ for $i = j$ and $\delta_{ij} = 0$ for $i \neq j$. One can see that the orthogonal vectors $|\alpha_i\rangle$, which satisfy condition (2.22), are linearly independent.

The *dimension* n of a vector space is given by the maximum number of linearly independent vectors . A set of linearly independent vectors $|\alpha_1\rangle, |\alpha_2\rangle, \ldots, |\alpha_n\rangle$ in an n-dimensional vector space is said to be a *basis* for the vector space. Since any vector $|\alpha\rangle$ can be expanded over a basis $\{|\alpha_1\rangle, |\alpha_2\rangle, \ldots, |\alpha_n\rangle\}$,

$$|\alpha\rangle = \sum_{i=1}^{n} a_i |\alpha_i\rangle, \tag{2.23}$$

we call the vectors $|\alpha_i\rangle$ a *complete set* of vectors. The complex numbers a_i are known as the components of the vector $|\alpha\rangle$ with respect to the basis $\{|\alpha_1\rangle, |\alpha_2\rangle, \ldots, |\alpha_n\rangle\}$. They are uniquely determined and, for an orthonormal basis, we have:

$$a_i = \langle \alpha_i | \alpha \rangle. \tag{2.24}$$

The ordered ensemble of components $\{a_1, a_2, \ldots, a_n\}$ constitutes a *representation* of the vector $|\alpha\rangle$.

An example of special interest for us is the vector space $\mathbf{C}^2$. A generic vector $|\alpha\rangle \in \mathbf{C}^2$ can be written as

$$|\alpha\rangle = a_1|\alpha_1\rangle + a_2|\alpha_2\rangle, \tag{2.25}$$

where the vectors $|\alpha_1\rangle$ and $|\alpha_2\rangle$ have components

$$\alpha_1 = (1, 0), \qquad \alpha_2 = (0, 1). \tag{2.26}$$

We shall use the following notation:

$$|\alpha_1\rangle = \begin{bmatrix} 1 \\ 0 \end{bmatrix}, \qquad |\alpha_2\rangle = \begin{bmatrix} 0 \\ 1 \end{bmatrix}. \tag{2.27}$$

This basis has nothing special. A generic vector $|\alpha\rangle$ can be expanded over any (orthonormal) basis. For instance, instead of Eq. (2.25) we can write

$$|\alpha\rangle = a_1'|\alpha_1'\rangle + a_2'|\alpha_2'\rangle, \tag{2.28}$$

where

$$|\alpha_1'\rangle = \frac{1}{\sqrt{2}} \begin{bmatrix} 1 \\ 1 \end{bmatrix}, \qquad |\alpha_2'\rangle = \frac{1}{\sqrt{2}} \begin{bmatrix} 1 \\ -1 \end{bmatrix}. \tag{2.29}$$

It is easy to check that the coefficients a_1' and a_2' are related to the coefficients a_1 and a_2 of the expansion (2.25) as follows:

$$a_1' = \frac{1}{\sqrt{2}}(a_1 + a_2), \qquad a_2' = \frac{1}{\sqrt{2}}(a_1 - a_2). \tag{2.30}$$

We now compute the inner product of two generic vectors $|\alpha\rangle$ and $|\beta\rangle$:

$$\langle\alpha|\beta\rangle = \left(\sum_i a_i^\star \langle\alpha_i|\right)\left(\sum_j b_j|\alpha_j\rangle\right) = \sum_{i,j} a_i^\star b_j \langle\alpha_i|\alpha_j\rangle = \sum_i a_i^\star b_i. \tag{2.31}$$

In particular, the norm of a generic vector $|\alpha\rangle$ can be written as

$$\||\alpha\rangle\| = \sqrt{\langle\alpha|\alpha\rangle} = \sqrt{\sum_i |a_i|^2}. \tag{2.32}$$

Note that the representation described above generalizes the expansion of a vector over orthogonal axes in three-dimensional Euclidean space. In particular, in this case the inner product becomes the usual scalar product of two vectors $\boldsymbol{u} = (u_1, u_2, u_3)$ and $\boldsymbol{v} = (v_1, v_2, v_3)$:

$$\boldsymbol{u} \cdot \boldsymbol{v} = |\boldsymbol{u}|\,|\boldsymbol{v}|\cos(\boldsymbol{u}, \boldsymbol{v}) = \sum_{i=1}^{3} u_i v_i \,, \tag{2.33}$$

where $(\boldsymbol{u}, \boldsymbol{v})$ is the angle between the vectors $\boldsymbol{u}$ and $\boldsymbol{v}$.

Gram–Schmidt decomposition

The Gram–Schmidt decomposition permits the construction of an orthonormal basis. Let us consider a basis $\{|\alpha_1\rangle, |\alpha_2\rangle, \ldots, |\alpha_n\rangle\}$ in an n-dimensional Hilbert space. It is easy to check that the vectors

$$|\beta_1\rangle = |\alpha_1\rangle \tag{2.34a}$$

and

$$|\beta_2\rangle = |\alpha_2\rangle - \frac{\langle\beta_1|\alpha_2\rangle}{\big\||\beta_1\rangle\big\|^2}|\beta_1\rangle \tag{2.34b}$$

are mutually orthogonal. We can define inductively, for any $i = 2, 3, \ldots, n$, the vector

$$|\beta_i\rangle = |\alpha_i\rangle - \sum_{k=1}^{i-1} \frac{\langle\beta_k|\alpha_i\rangle}{\big\||\beta_k\rangle\big\|^2}|\beta_k\rangle \,. \tag{2.34c}$$

It is easy to see that the vectors $\{|\beta_1\rangle, |\beta_2\rangle, \ldots, |\beta_n\rangle\}$ are mutually orthogonal. Thus, an orthonormal basis for the Hilbert space is given by

$$|\gamma_i\rangle = \frac{|\beta_i\rangle}{\big\||\beta_i\rangle\big\|} \,, \qquad (i = 1, 2, \ldots, n) \,. \tag{2.35}$$

Linear operators

An operator A maps each vector $|\alpha\rangle \in V$ into another vector $|\beta\rangle \in V$:

$$|\beta\rangle = A|\alpha\rangle \,. \tag{2.36}$$

The operator A is said to be *linear* if, for any vectors $|\alpha\rangle$ and $|\beta\rangle$ and for any complex numbers a and b, the following property holds:

$$A\big(a|\alpha\rangle + b|\beta\rangle\big) = aA|\alpha\rangle + bA|\beta\rangle \,. \tag{2.37}$$

The simplest example of a linear operator is the *identity* operator I:

$$I|\alpha\rangle = |\alpha\rangle \,. \tag{2.38}$$

Another simple example is the *zero* operator N, which maps any vector $|\alpha\rangle \in V$ into the zero vector 0:

$$N|\alpha\rangle = 0. \tag{2.39}$$

Two operators A and B are said to be equal (and we write $A = B$) if, for any vector $|\alpha\rangle \in V$,

$$A|\alpha\rangle = B|\alpha\rangle. \tag{2.40}$$

The sum $C = A + B$ of two linear operators A and B is linear and is defined as follows:

$$C\,|\alpha\rangle = (A + B)\,|\alpha\rangle = A\,|\alpha\rangle + B\,|\alpha\rangle. \tag{2.41}$$

We define the product $D = AB$ of two operators by means of the relation

$$D\,|\alpha\rangle = AB\,|\alpha\rangle = A\big(B\,|\alpha\rangle\big). \tag{2.42}$$

Therefore, the application of the operator $D = AB$ to the vector $|\alpha\rangle$ is equivalent to first applying B to $|\alpha\rangle$ and then A to the vector $B|\alpha\rangle$. While it is easy to check that $A + B = B + A$, in general $AB \neq BA$. As we shall see later, it is only in special cases that the two operators commute, that is, $AB = BA$.

Completeness relation

Starting from the relation $a_i = \langle\alpha_i|\alpha\rangle$ $(i = 1, \ldots, n)$, we obtain

$$\Big(\sum_i |\alpha_i\rangle\langle\alpha_i|\Big)|\alpha\rangle = \sum_i |\alpha_i\rangle\langle\alpha_i|\alpha\rangle = \sum_i a_i|\alpha_i\rangle = |\alpha\rangle. \tag{2.43}$$

Note that $\sum_i |\alpha_i\rangle\langle\alpha_i|$ is an operator, since it maps a vector into a vector. Since relation (2.43) applies for any vector $|\alpha\rangle$, we have the completeness relation

$$\sum_i |\alpha_i\rangle\langle\alpha_i| = I, \tag{2.44}$$

where I is the identity operator defined by Eq. (2.38).

Matrix representation

A linear operator A can be represented as a square matrix by means of a complete set of vectors. Let us consider the action of the operator A on a generic vector $|\alpha\rangle \in V$, namely, $A|\alpha\rangle = |\beta\rangle$. We expand the two vectors $|\alpha\rangle$ and $|\beta\rangle$ over an orthonormal basis $\{|\gamma_1\rangle, |\gamma_2\rangle, \ldots, |\gamma_n\rangle\}$:

$$|\alpha\rangle = \sum_i a_i|\gamma_i\rangle, \qquad |\beta\rangle = \sum_i b_i|\gamma_i\rangle, \qquad (2.45)$$

and therefore

$$b_i = \langle\gamma_i|\beta\rangle = \langle\gamma_i|A\alpha\rangle = \sum_j \langle\gamma_i|A\gamma_j\rangle a_j \equiv \sum_j A_{ij}a_j \quad (i = 1, 2, \ldots, n),$$
$$(2.46)$$

where we have defined

$$A_{ij} = \langle\gamma_i|A\gamma_j\rangle. \qquad (2.47)$$

Note that we shall also use the notation $\langle\gamma_i|A|\gamma_j\rangle \equiv \langle\gamma_i|A\gamma_j\rangle$. The system of equations (2.46) reads

$$
\begin{bmatrix} b_1 \\ b_2 \\ \vdots \\ b_n \end{bmatrix}
=
\begin{bmatrix} A_{11} & A_{12} & \ldots & A_{1n} \\ A_{21} & A_{22} & \ldots & A_{2n} \\ \vdots & \vdots & & \vdots \\ A_{n1} & A_{n2} & \ldots & A_{nn} \end{bmatrix}
\begin{bmatrix} a_1 \\ a_2 \\ \vdots \\ a_n \end{bmatrix},
\qquad (2.48)
$$

where a generic vector $|\alpha\rangle$ is represented as a column vector:

$$|\alpha\rangle = \begin{bmatrix} a_1 \\ a_2 \\ \vdots \\ a_n \end{bmatrix}. \qquad (2.49)$$

Thus, if we know all the matrix elements A_{ij}, we can compute the action of the operator A on a generic vector $|\alpha\rangle \in V$, using relation (2.48). Note that it is also possible to represent the inner product as follows:

$$\langle\alpha|\beta\rangle = \sum_i a_i^\star b_i = [a_1^\star, a_2^\star, \ldots, a_n^\star] \begin{bmatrix} b_1 \\ b_2 \\ \vdots \\ b_n \end{bmatrix}. \qquad (2.50)$$

Pauli matrices

In this book, we shall often use the Pauli matrices σ_x, σ_y and σ_z, defined as follows:

$$\sigma_x = \begin{bmatrix} 0 & 1 \\ 1 & 0 \end{bmatrix}, \qquad \sigma_y = \begin{bmatrix} 0 & -i \\ i & 0 \end{bmatrix}, \qquad \sigma_z = \begin{bmatrix} 1 & 0 \\ 0 & -1 \end{bmatrix}. \qquad (2.51)$$

These matrices have the following relevant properties:

(i) $\sigma_x^2 = \sigma_y^2 = \sigma_z^2 = I$, where I is the identity matrix

$$I = \begin{bmatrix} 1 & 0 \\ 0 & 1 \end{bmatrix}; \qquad (2.52)$$

(ii) $\sigma_x \sigma_y = i\sigma_z$, $\sigma_y \sigma_z = i\sigma_x$, $\sigma_z \sigma_x = i\sigma_y$.

Projectors

An important class of operators is given by the *projectors*. If $|\alpha\rangle \in V$ is a unit vector, the unidimensional projector P_α is defined, for any vector $|\gamma\rangle \in V$, as follows:

$$|\beta\rangle = P_\alpha|\gamma\rangle = |\alpha\rangle\langle\alpha|\gamma\rangle = \langle\alpha|\gamma\rangle|\alpha\rangle. \qquad (2.53)$$

This operator is called a projector since it projects a generic vector $|\gamma\rangle$ along the direction $|\alpha\rangle$. In particular, $P_\alpha|\alpha\rangle = |\alpha\rangle$ and $P_\alpha|\gamma\rangle = 0$ for any $|\gamma\rangle$ orthogonal to $|\alpha\rangle$. A projector satisfies the following property:

$$P_\alpha^2 = P_\alpha. \qquad (2.54)$$

This property is easy to check by taking into account that $P_\alpha|\alpha\rangle = |\alpha\rangle$.

Definition (2.53) is readily extended to projectors over multi-dimensional subspaces. We have

$$P = \sum_{l=1}^{k} |\alpha_l\rangle\langle\alpha_l|, \qquad (2.55)$$

where k is the dimension of the subspace over which the operator P projects. Again, it is easy to check that $P^2 = P$. We note that it is also possible to prove that a linear operator P, such that $P^2 = P$, is a projector; therefore this property can be taken as the definition of a projector.

Eigenvalues and eigenvectors

An *eigenvector* of a linear operator A is a non-zero vector $|\alpha\rangle$ such that

$$A|\alpha\rangle = \alpha|\alpha\rangle, \qquad (2.56)$$

where α is a complex number called the *eigenvalue* of A corresponding to the eigenvector $|\alpha\rangle$. The eigenvalue equation (2.56) always has a solution. Indeed, we can expand the vectors $|\alpha\rangle$ and $A|\alpha\rangle$ over an orthonormal basis $\{|\gamma_1\rangle, |\gamma_2\rangle, \ldots, |\gamma_n\rangle\}$ as follows:

$$|\alpha\rangle = \sum_{i=1}^{n} a_i|\gamma_i\rangle \qquad \left(a_i \equiv \langle\gamma_i|\alpha\rangle\right), \qquad (2.57)$$

$$A|\alpha\rangle = \sum_{i=1}^{n} c_i|\gamma_i\rangle, \qquad (2.58)$$

where

$$c_i = \langle\gamma_i|A|\alpha\rangle = \sum_j \langle\gamma_i|A|\gamma_j\rangle a_j = \sum_j A_{ij} a_j. \qquad (2.59)$$

If we insert these expansions into Eq. (2.56) we obtain

$$\sum_{i=1}^{n} \left(\sum_{j=1}^{n} A_{ij} a_j - \alpha a_i\right)|\gamma_i\rangle = 0, \qquad (2.60)$$

which is satisfied if and only if

$$\sum_{j=1}^{n} A_{ij} a_j - \alpha a_i = \sum_{j=1}^{n}(A_{ij} - \alpha\delta_{ij}) a_j = 0, \quad (i = 1, 2, \ldots, n). \qquad (2.61)$$

This system of homogeneous linear equations has non-zero solutions if and only if the eigenvalue α satisfies the *characteristic equation*

$$\det(A - \alpha I) = \det \begin{bmatrix} A_{11} - \alpha & A_{12} & \ldots & A_{1n} \\ A_{21} & A_{22} - \alpha & \ldots & A_{2n} \\ \vdots & \vdots & & \vdots \\ A_{n1} & A_{n2} & \ldots & A_{nn} - \alpha \end{bmatrix} = 0. \qquad (2.62)$$

The solutions to the characteristic equation are the eigenvalues of the linear operator A. Since $p(\alpha) \equiv \det(A - \alpha I)$ is a polynomial of degree n, a fundamental theorem of algebra tells us that the equation $p(\alpha) = 0$ has n complex roots (eigenvalues) $\alpha_1, \alpha_2, \ldots, \alpha_n$. This shows that Eq. (2.56) always has a solution. It is possible to prove that the characteristic equation depends only on the operator A and not on the specific matrix representation used for A. Therefore, the eigenvalues of a linear operator are independent of its matrix representation.

Exercise 2.1 Show that the eigenvectors of a linear operator A belonging to distinct eigenvalues are linearly independent.

Hermitian operators

For any linear operator A on a Hilbert space $\mathcal{H}$, it is possible to show that there exists a unique linear operator $A^\dagger$ on $\mathcal{H}$, called the *adjoint* or *Hermitian conjugate* of A, such that, for all vectors $|\alpha\rangle, |\beta\rangle \in \mathcal{H}$,

$$\langle \alpha | A\beta \rangle = \langle A^\dagger \alpha | \beta \rangle. \tag{2.63}$$

Starting from the definition (2.63), it is easy to see that $\langle A\alpha | \beta \rangle = \langle \alpha | A^\dagger \beta \rangle$. (Indeed, $\langle A\alpha | \beta \rangle = \langle \beta | A\alpha \rangle^\star = \langle A^\dagger \beta | \alpha \rangle^\star = \langle \alpha | A^\dagger \beta \rangle$.)

Exercise 2.2 Show that $(A+B)^\dagger = A^\dagger + B^\dagger$, $(AB)^\dagger = B^\dagger A^\dagger$ and $(A^\dagger)^\dagger = A$.

A particularly interesting case is that in which A is *Hermitian* or *self-adjoint*; that is, it is equal to its own adjoint:

$$A^\dagger = A. \tag{2.64}$$

In this case, the scalar product $\langle \alpha | A\alpha \rangle$ is real (since $\langle \alpha | A\alpha \rangle^\star = \langle A\alpha | \alpha \rangle = \langle \alpha | A\alpha \rangle$). This implies that the eigenvalues of an Hermitian operator are real. Indeed, if $A|\alpha\rangle = \alpha|\alpha\rangle$, then $\langle \alpha | A\alpha \rangle = \alpha\langle \alpha | \alpha \rangle$ and, since both $\langle \alpha | A\alpha \rangle$ and $\langle \alpha | \alpha \rangle$ are real, the eigenvalue α also has to be real.

The eigenvectors of an Hermitian operator form an orthonormal set in the Hilbert space $\mathcal{H}$. (It is assumed that the eigenvectors have unit norm; if not, they can be normalized by dividing them by their norms.) This property is easy to prove: assume that α_1 and α_2 are distinct eigenvalues corresponding to the eigenvectors $|\alpha_1\rangle$ and $|\alpha_2\rangle$; we have

$$\langle \alpha_j | A\alpha_i \rangle = \alpha_i \langle \alpha_j | \alpha_i \rangle, \tag{2.65a}$$

$$\langle A\alpha_j | \alpha_i \rangle = \alpha_j \langle \alpha_j | \alpha_i \rangle. \tag{2.65b}$$

Subtracting side-by-side (2.65b) from (2.65a), we obtain $(\alpha_i - \alpha_j)\langle \alpha_j | \alpha_i \rangle = 0$ and, since $\alpha_i \neq \alpha_j$, we obtain $\langle \alpha_j | \alpha_i \rangle = 0$. Here we have assumed that the eigenvalues α_i are not degenerate, that is, $\alpha_i \neq \alpha_j$ for $i \neq j$. However, it can be shown that it is also possible to construct an orthonormal set of eigenvectors of the operator A in the degenerate case, in which there are linearly independent eigenvectors corresponding to the same eigenvalue. In summary, given an Hermitian operator A, it is always possible to construct an orthonormal basis of eigenvectors of A. Therefore, any vector in the Hilbert space $\mathcal{H}$ can be expressed as a linear superposition of vectors of this basis. This property is called *completeness* and the basis of eigenvectors of A is said to be a *complete orthonormal set*.

Let us consider the matrix representation of a linear operator A over a basis $\{|\gamma_1\rangle, |\gamma_2\rangle, \ldots, |\gamma_n\rangle\}$:

$$A_{ij} \equiv \langle \gamma_i | A \gamma_j \rangle. \tag{2.66}$$

From the definition of an adjoint operator, we have that $\langle A\gamma_i | \gamma_j \rangle = \langle \gamma_i | A^\dagger \gamma_j \rangle$ and this relation can be written as

$$(A_{ji})^\star = (A^\dagger)_{ij}. \tag{2.67}$$

Therefore, the matrix elements of $A^\dagger$ are the complex conjugates of the matrix elements of the *transpose matrix* A^T:

$$A^\dagger = (A^T)^\star. \tag{2.68}$$

(The transpose matrix is defined by $A^T_{ij} = A_{ji}$.) For an Hermitian operator, we have

$$A = (A^T)^\star. \tag{2.69}$$

As a consequence, the diagonal matrix elements of an Hermitian operator are real: $A_{ii} = (A^T_{ii})^\star = (A_{ii})^\star$.

Inverse operator

Let us consider a linear operator A. If there exists an operator B such that

$$AB = BA = I, \tag{2.70}$$

we call B the inverse of A and write $B = A^{-1}$. If we have $|\beta\rangle = A|\alpha\rangle$, then $|\alpha\rangle = A^{-1}|\beta\rangle$. It is possible to show that the inverse of an operator A exists if and only if the equation $A|\alpha\rangle = 0$ implies that $|\alpha\rangle$ is the zero vector. Considering the matrix representation of A, it is immediate to conclude that the inverse of an operator A exists if and only if

$$\det A \neq 0. \tag{2.71}$$

Exercise 2.3 Show that a projector P is Hermitian and can be inverted if and only if $P = I$.

Unitary operators

An operator U is said to be unitary if

$$UU^\dagger = U^\dagger U = I. \tag{2.72}$$

From this definition, we have that the adjoint of a unitary operator coincides with its inverse,

$$U^\dagger = U^{-1}, \tag{2.73}$$

and that $U^\dagger$ is unitary. The product UV of two unitary operators is unitary, since

$$(UV)(UV)^\dagger = UVV^\dagger U^\dagger = I. \tag{2.74}$$

Unitary operators have the important property that they preserve the inner product between vectors. To see this, let us consider any two vectors $|\alpha\rangle$ and $|\beta\rangle$. If we define $|\gamma\rangle = U|\alpha\rangle$ and $|\nu\rangle = U|\beta\rangle$, then

$$\langle\gamma|\nu\rangle = \langle U\alpha|U\beta\rangle = \langle\alpha|U^\dagger U|\beta\rangle = \langle\alpha|\beta\rangle. \tag{2.75}$$

If we take $|\alpha\rangle = |\beta\rangle$, we see that a unitary operator does not change the norm of a vector. Therefore, unitary operators act on vectors in Hilbert space in a way analogous to rotations in Euclidean space, which preserve both the length of a vector and the angle between two vectors.

Exercise 2.4 Show that the Pauli matrices σ_x, σ_y and σ_z, defined by Eq. (2.51), are both Hermitian and unitary.

Change of basis

It is possible to change representation, namely, to pass from an orthonormal basis $(|\gamma_i\rangle)$ to another $(|\gamma_i'\rangle)$ by means of a unitary transformation S:

$$|\gamma_i'\rangle = \sum_j S_{ji}|\gamma_j\rangle \qquad (i = 1, 2, \ldots, n). \tag{2.76}$$

A generic vector

$$|\alpha\rangle = \sum_i a_i|\gamma_i\rangle \qquad \left(a_i \equiv \langle\gamma_i|\alpha\rangle\right), \tag{2.77}$$

can be expressed in the new basis as

$$|\alpha\rangle = \sum_j a_j'|\gamma_j'\rangle = \sum_{ij} a_j' S_{ij}|\gamma_i\rangle \qquad \left(a_j' \equiv \langle\gamma_j'|\alpha\rangle\right), \tag{2.78}$$

where we have used Eq. (2.76). Thus, the old and new vector components are linked by the relation

$$a_i = \sum_j S_{ij}\, a'_j\,. \tag{2.79}$$

Exercise 2.5 Show that the matrix representations A and A' of an operator A with respect to the bases $|\gamma_i\rangle$ and $|\gamma'_i\rangle$ are connected as follows:

$$A' = S^{-1}AS\,. \tag{2.80}$$

A very important representation of an operator A is its *diagonal representation*, in which the basis is given by the eigenvectors of A. With respect to this basis, the matrix representation of A reads as follows:

$$A = \sum_{i=1}^{n} \lambda_i\, |i\rangle\langle i|\,, \tag{2.81}$$

where λ_i are the eigenvalues of A and $|i\rangle$ the corresponding eigenvectors. We call Eq. (2.81) the *spectral decomposition* of the operator A (the ensemble of the eigenvalues of A constitutes its *spectrum*).

An example of a diagonal representation is the Pauli matrix

$$\sigma_z = \begin{bmatrix} 1 & 0 \\ 0 & -1 \end{bmatrix} = |0\rangle\langle 0| - |1\rangle\langle 1|\,, \tag{2.82}$$

which is diagonal with respect to the eigenvector basis

$$|0\rangle = \begin{bmatrix} 1 \\ 0 \end{bmatrix}, \qquad |1\rangle = \begin{bmatrix} 0 \\ 1 \end{bmatrix}, \tag{2.83}$$

where the eigenvectors $|0\rangle$ and $|1\rangle$ correspond to the eigenvalues $+1$ and -1, respectively.

In the representation $\{|0\rangle, |1\rangle\}$, the Pauli matrix σ_x reads

$$\sigma_x = \begin{bmatrix} 0 & 1 \\ 1 & 0 \end{bmatrix} = |0\rangle\langle 1| + |1\rangle\langle 0|\,. \tag{2.84}$$

The operator σ_x is diagonal in the basis

$$|+\rangle = \frac{1}{\sqrt{2}} \begin{bmatrix} 1 \\ 1 \end{bmatrix}, \qquad |-\rangle = \frac{1}{\sqrt{2}} \begin{bmatrix} 1 \\ -1 \end{bmatrix}, \tag{2.85}$$

in which its matrix representation is given by

$$\sigma_x = |+\rangle\langle +| - |-\rangle\langle -|\,. \tag{2.86}$$

The new basis $\{|+\rangle, |-\rangle\}$ is related to the original basis $\{|0\rangle, |1\rangle\}$ by means of the unitary transformation

$$S = \frac{1}{\sqrt{2}} \begin{bmatrix} 1 & 1 \\ 1 & -1 \end{bmatrix}. \tag{2.87}$$

Exercise 2.6   Write down the Pauli matrices in the basis $\{|+\rangle, |-\rangle\}$.

An operator is said to be *diagonalizable* if it has a diagonal representation. There are operators that are not diagonalizable, such as the operator with matrix representation

$$\begin{bmatrix} 1 & 1 \\ 0 & 1 \end{bmatrix}. \tag{2.88}$$

In this example there is only one eigenvalue, $\lambda = 1$, and the corresponding eigenvector

$$|u\rangle = \begin{bmatrix} 1 \\ 0 \end{bmatrix} \tag{2.89}$$

spans a one-dimensional subspace and therefore cannot be a basis for the two-dimensional vector space on which the matrix (2.88) operates. It is possible to show that both Hermitian and unitary operators are diagonalizable. Indeed, these operators belong to the larger class of *normal operators*, defined by the condition

$$AA^\dagger = A^\dagger A. \tag{2.90}$$

We state without proof the following remarkable theorem:

Theorem 2.1   Spectral decomposition theorem: *An operator is diagonalizable, with orthonormal basis of eigenvectors, if and only if it is normal.*

Commutators

We say that two operators A and B commute if they satisfy the following relation:

$$AB = BA. \tag{2.91}$$

The *commutator* of two operators A and B is defined by

$$[A, B] = AB - BA. \tag{2.92}$$

It is easy to check the following properties:

$$[A, B] = -[B, A], \tag{2.93a}$$

$$[AB, C] = A[B, C] + [A, C]B. \tag{2.93b}$$

Exercise 2.7   Show that, if A and B are Hermitian, $i[A, B]$ is Hermitian.

Theorem 2.2   <u>Simultaneous diagonalization theorem</u>: *Two normal operators A and B commute if and only if there exists an orthonormal basis with respect to which both A and B are diagonal.*

Proof.   Assume that $|i\rangle$ is an orthonormal basis for both A and B, that is,

$$A|i\rangle = \lambda_i |i\rangle, \qquad B|i\rangle = \nu_i |i\rangle. \tag{2.94}$$

Therefore,

$$AB|i\rangle = A\nu_i|i\rangle = \lambda_i \nu_i|i\rangle = \nu_i \lambda_i|i\rangle = BA|i\rangle. \tag{2.95}$$

Thus, $[A, B] = 0$. To show the converse, let us denote by $|i\rangle$ an orthonormal basis for the operator A with eigenvalues λ_i. Assume that the vectors $|i\rangle$ are not eigenfunctions of the operator B and expand $B|i\rangle$ on the $|i\rangle$ basis:

$$B|i\rangle = \sum_{j=1}^{n} \langle j|B|i\rangle|j\rangle. \tag{2.96}$$

Therefore,

$$
\begin{aligned}
[A, B]|i\rangle = AB|i\rangle - BA|i\rangle &= \sum_{j=1}^{n} \langle j|B|i\rangle\lambda_j|j\rangle - \lambda_i \sum_{j=1}^{n} \langle j|B|i\rangle|j\rangle \\
&= \sum_{j=1}^{n} \langle j|B|i\rangle(\lambda_j - \lambda_i)|j\rangle = 0.
\end{aligned} \tag{2.97}
$$

Here we have $[A, B]|i\rangle = 0$, since we assume that A and B commute. If the eigenvalues are not degenerate, that is, $\lambda_i \neq \lambda_j$ for $i \neq j$, then from Eq. (2.97) we obtain

$$\langle j|B|i\rangle = 0 \qquad \text{for} \quad i \neq j. \tag{2.98}$$

If we denote $\langle j|B|j\rangle = \nu_j$, we have

$$\langle j|B|i\rangle = \nu_j \, \delta_{ij}. \tag{2.99}$$

Inserting this relation into Eq. (2.96), we obtain

$$B|i\rangle = \nu_i |i\rangle. \tag{2.100}$$

Therefore, $|i\rangle$ is also an eigenvector of the operator B. Note that the proof can be extended to the case in which there are degeneracies in the eigenvalues λ_i. $\square$

The anti-commutator

The anti-commutator of two operators A and B is defined by

$$\{A, B\} = AB + BA. \tag{2.101}$$

We say that two operators A and B *anti-commute* if $\{A, B\} = 0$.

It is easy to verify that the Pauli matrices anti-commute,

$$\{\sigma_i, \sigma_j\} = 0, \qquad (i, j = x, y, z), \tag{2.102}$$

while the following commutation relations hold:

$$[\sigma_x, \sigma_y] = 2i\sigma_z, \qquad [\sigma_y, \sigma_z] = 2i\sigma_x, \qquad [\sigma_z, \sigma_x] = 2i\sigma_y. \tag{2.103}$$

Trace

The trace of a matrix A is defined as the sum of its diagonal elements:

$$\mathrm{Tr}(A) = \sum_{i=1}^{n} A_{ii}. \tag{2.104}$$

It is easy to check the following properties:

(i) $\mathrm{Tr}(A + B) = \mathrm{Tr}(A) + \mathrm{Tr}(B)$ (linearity); $\tag{2.105a}$

(ii) $\mathrm{Tr}(cA) = c\,\mathrm{Tr}(A)$ $(c \in C)$; $\tag{2.105b}$

(iii) $\mathrm{Tr}(AB) = \mathrm{Tr}(BA)$ (cyclic property). $\tag{2.105c}$

Note that, as a consequence of the cyclic property, for n operators A_1, A_2, $\ldots$, A_n, we have

$$\mathrm{Tr}(A_1 A_2 \cdots A_{n-1} A_n) = \mathrm{Tr}(A_2 A_3 \cdots A_n A_1) = \ldots$$
$$\ldots = \mathrm{Tr}(A_n A_1 \cdots A_{n-2} A_{n-1}). \tag{2.106}$$

The trace of an operator A is defined as the trace of a matrix representation of A. It is easy to check that the trace is independent of the choice of representation. Indeed, from the relation $\sum_j |j\rangle\langle j| = I$, we obtain

$$\mathrm{Tr}(A) = \sum_i \langle i|A|i\rangle = \sum_i \sum_j \sum_k \langle i|j\rangle\langle j|A|k\rangle\langle k|i\rangle$$
$$= \sum_j \sum_k \langle k|j\rangle\langle j|A|k\rangle = \sum_j \langle j|A|j\rangle. \tag{2.107}$$

From property (iii), it follows that, for a unitary operator U,

$$\mathrm{Tr}(U^\dagger A U) = \mathrm{Tr}(U U^\dagger A) = \mathrm{Tr}(A), \tag{2.108}$$

and therefore the trace is invariant under unitary transformations. Another important property, which we shall use later, is the following: let $|i\rangle$ be an orthonormal basis and $|\alpha\rangle$ a generic vector, which we can expand over the $|i\rangle$ basis as follows: $|\alpha\rangle = \sum_i \langle i|\alpha\rangle |i\rangle$. Then

$$\text{Tr}(A|\alpha\rangle\langle\alpha|) = \sum_i \langle i|A|\alpha\rangle\langle\alpha|i\rangle = \sum_i \langle\alpha|i\rangle\langle i|A|\alpha\rangle = \langle\alpha|A|\alpha\rangle. \quad (2.109)$$

Tensor product

Let us consider two Hilbert spaces $\mathcal{H}_1$ and $\mathcal{H}_2$ of dimension m and n, respectively. We say that the Hilbert space $\mathcal{H}$ is the tensor product of $\mathcal{H}_1$ and $\mathcal{H}_2$, and we write $\mathcal{H} = \mathcal{H}_1 \otimes \mathcal{H}_2$, if we can associate with each pair of vectors $|\alpha\rangle \in \mathcal{H}_1$ and $|\beta\rangle \in \mathcal{H}_2$ a vector belonging to $\mathcal{H}$, denoted by $|\alpha\rangle \otimes |\beta\rangle$ and called the tensor product of $|\alpha\rangle$ and $|\beta\rangle$. By definition, the vectors in $\mathcal{H}$ are linear superpositions of the above vectors $|\alpha\rangle \otimes |\beta\rangle$ and the following properties are satisfied:

(i) for any $|\alpha\rangle \in \mathcal{H}_1$, $|\beta\rangle \in \mathcal{H}_2$ and $c \in \mathbf{C}$,

$$c\big(|\alpha\rangle \otimes |\beta\rangle\big) = \big(c|\alpha\rangle\big) \otimes |\beta\rangle = |\alpha\rangle \otimes \big(c|\beta\rangle\big); \quad (2.110a)$$

(ii) for any $|\alpha_1\rangle, |\alpha_2\rangle \in \mathcal{H}_1$ and $|\beta\rangle \in \mathcal{H}_2$,

$$\big(|\alpha_1\rangle + |\alpha_2\rangle\big) \otimes |\beta\rangle = |\alpha_1\rangle \otimes |\beta\rangle + |\alpha_2\rangle \otimes |\beta\rangle; \quad (2.110b)$$

(iii) for any $|\alpha\rangle \in \mathcal{H}_1$ and $|\beta_1\rangle, |\beta_2\rangle \in \mathcal{H}_2$,

$$|\alpha\rangle \otimes \big(|\beta_1\rangle + |\beta_2\rangle\big) = |\alpha\rangle \otimes |\beta_1\rangle + |\alpha\rangle \otimes |\beta_2\rangle. \quad (2.110c)$$

Note that, instead of $|\alpha\rangle \otimes |\beta\rangle$, we shall often use the shorthand notations $|\alpha\rangle|\beta\rangle$, $|\alpha, \beta\rangle$ or $|\alpha\beta\rangle$.

The dimension of the Hilbert space $\mathcal{H}$ is given by the product mn of the dimensions of $\mathcal{H}_1$ and $\mathcal{H}_2$. Indeed, if $|i\rangle$ and $|j\rangle$ are orthonormal bases for $\mathcal{H}_1$ and $\mathcal{H}_2$, then an orthonormal basis for $\mathcal{H} = \mathcal{H}_1 \otimes \mathcal{H}_2$ is given by $|i\rangle \otimes |j\rangle$. For instance, if $\mathcal{H}_1$ and $\mathcal{H}_2$ are two-dimensional Hilbert spaces ($m = n = 2$) with basis vectors $|0\rangle$ and $|1\rangle$, then $\mathcal{H}$ has dimension $mn = 4$ and basis vectors $|0\rangle \otimes |0\rangle$, $|0\rangle \otimes |1\rangle$, $|1\rangle \otimes |0\rangle$ and $|1\rangle \otimes |1\rangle$. Therefore, a generic vector $|\Psi\rangle \in \mathcal{H}$ can be expanded over this basis as follows:

$$|\Psi\rangle = c_{00}|00\rangle + c_{01}|01\rangle + c_{10}|10\rangle + c_{11}|11\rangle, \quad (2.111)$$

where $c_{ij} \equiv \langle ij|\Psi\rangle$, with $i, j = 0, 1$.

If A and B are linear operators acting on $\mathcal{H}_1$ and $\mathcal{H}_2$, respectively, then the action of $A \otimes B$ on a generic vector

$$|\Psi\rangle = \sum_{ij} c_{ij} |i\rangle \otimes |j\rangle \qquad (2.112)$$

residing in $\mathcal{H}$ is defined by

$$(A \otimes B)\left(\sum_{ij} c_{ij} |i\rangle \otimes |j\rangle\right) = \sum_{ij} c_{ij} A |i\rangle \otimes B |j\rangle. \qquad (2.113)$$

It is possible to show that a generic linear operator O acting on $\mathcal{H}$ can be written as a linear superposition of tensor products of linear operators A_i acting on $\mathcal{H}_1$ and B_j acting on $\mathcal{H}_2$:

$$O = \sum_{ij} \gamma_{ij} A_i \otimes B_j. \qquad (2.114)$$

The inner product of two vectors $|\Psi\rangle$ and $|\Phi\rangle \in \mathcal{H}$, with $|\Psi\rangle = \sum_{ij} c_{ij} |ij\rangle$ and $|\Phi\rangle = \sum_{ij} d_{ij} |ij\rangle$, is defined by

$$\langle\Psi|\Phi\rangle = \sum_{ij} c_{ij}^{\star} d_{ij}. \qquad (2.115)$$

It is easy to show that this definition satisfies the properties of an inner product.

The matrix representation of the operator $A \otimes B$ in the basis $|K\rangle \equiv |ij\rangle$, labelled by the single index $K = 1, 2, \ldots, mn$, with $K = (i-1)n + j$, is given by

$$A \otimes B = \begin{bmatrix} A_{11}B & A_{12}B & \cdots & A_{1m}B \\ A_{21}B & A_{22}B & \cdots & A_{2m}B \\ \vdots & \vdots & & \vdots \\ A_{m1}B & A_{m2}B & \cdots & A_{mm}B \end{bmatrix}, \qquad (2.116)$$

where the terms $A_{ij}B$ denote sub-matrices of size $n \times n$, with A and B matrix representations of the operators A and B (A and B are $m \times m$ and $n \times n$ matrices, respectively). For instance, let us consider the matrix representation of the tensor product of the Pauli matrices σ_x and σ_z:

$$\sigma_x \otimes \sigma_z = \begin{bmatrix} 0 & 1 \\ 1 & 0 \end{bmatrix} \otimes \begin{bmatrix} 1 & 0 \\ 0 & -1 \end{bmatrix} = \begin{bmatrix} 0 \cdot \sigma_z & 1 \cdot \sigma_z \\ 1 \cdot \sigma_z & 0 \cdot \sigma_z \end{bmatrix} = \begin{bmatrix} 0 & 0 & 1 & 0 \\ 0 & 0 & 0 & -1 \\ 1 & 0 & 0 & 0 \\ 0 & -1 & 0 & 0 \end{bmatrix}.$$

$$(2.117)$$

Exercise 2.8 Compute the tensor products $\sigma_x \otimes \sigma_y$ and $I \otimes \sigma_x$.

As a further example, let us consider the vectors $|\alpha\rangle = \frac{1}{\sqrt{2}}(|0\rangle - |1\rangle)$ and $|\beta\rangle = \frac{1}{\sqrt{2}}(|0\rangle + |1\rangle)$, and compute their tensor product $|\alpha\rangle \otimes |\beta\rangle$. It has matrix representation, with respect to the basis $\{|00\rangle, |01\rangle, |10\rangle, |11\rangle\}$, given by

$$|\alpha\rangle \otimes |\beta\rangle = \frac{1}{\sqrt{2}} \begin{bmatrix} 1 \cdot |\beta\rangle \\ -1 \cdot |\beta\rangle \end{bmatrix} = \frac{1}{2} \begin{bmatrix} 1 \\ 1 \\ -1 \\ -1 \end{bmatrix}. \qquad (2.118)$$

2.4 The postulates of quantum mechanics

In classical mechanics the state of a system of n particles at time t_0 is determined by the positions $\{x_1(t_0), x_2(t_0), \ldots, x_n(t_0)\}$ and the velocities $\{\dot{x}_1(t_0), \dot{x}_2(t_0), \ldots, \dot{x}_n(t_0)\}$ of all the particles at this time. If these initial conditions are known, Newton's laws of classical mechanics allow us, at least in principle, to compute the state of the system at any time t. Indeed, the laws of classical mechanics lead to first-order ordinary differential equations in the variables x_i and $\dot{x}_i$ and, once the initial conditions are set, there exists a unique solution $\{x_1(t), x_2(t), \ldots, x_n(t); \dot{x}_1(t), \dot{x}_2(t), \ldots, \dot{x}_n(t)\}$.

Quantum mechanics is based on a completely different mathematical framework. In the following, we shall introduce the postulates that are at the basis of quantum theory.

Postulate I: *The state of a physical system S is completely described by a unit vector $|\psi\rangle$, which is called the state vector, or wave function, and resides in the Hilbert space $\mathcal{H}_S$ associated with the system.*

The evolution in time of the state vector $|\psi\rangle$ is governed by the

Schrödinger equation

$$i\hbar\frac{d}{dt}|\psi(t)\rangle = H|\psi(t)\rangle, \tag{2.119}$$

where H is a self-adjoint operator known as the Hamiltonian of the system and $\hbar \equiv h/2\pi$, with the physical constant h known as Planck's constant. Its value ($h \approx 6.626\times10^{-34}$ Joule sec) is determined experimentally.

It is important to note that the Schrödinger equation is a linear differential equation of first order in time. Therefore, given the initial state $|\psi(t_0)\rangle$, the state $|\psi(t)\rangle$ at any time t is completely and uniquely determined by the solution to the Schrödinger equation.

Since the Schrödinger equation is linear, the following *superposition principle* applies: if $|\psi_1(t)\rangle$ and $|\psi_2(t)\rangle$ are solutions of Eq. (2.119), then the superposition $|\psi(t)\rangle = \alpha|\psi_1(t)\rangle + \beta|\psi_2(t)\rangle$, where α and β are complex numbers, is also a solution. Therefore, the *time-evolution operator U*, defined by

$$|\psi(t)\rangle = U(t, t_0)|\psi(t_0)\rangle \tag{2.120}$$

is linear. If the Hamiltonian H is time independent, the solution to the Schrödinger equation (2.119) can be written as

$$|\psi(t)\rangle = \exp\left[-\frac{i}{\hbar}H(t - t_0)\right]|\psi(t_0)\rangle, \tag{2.121}$$

and therefore

$$U(t, t_0) = \exp\left[-\frac{i}{\hbar}H(t - t_0)\right], \tag{2.122}$$

where the exponential of the operator $-iH(t - t_0)/\hbar$ is defined as follows:

$$\exp\left[-\frac{i}{\hbar}H(t - t_0)\right] \equiv \sum_{n=0}^{\infty}\frac{1}{n!}\left[-\frac{i}{\hbar}(t - t_0)\right]^n H^n. \tag{2.123}$$

Starting from this equation, it is easy to prove that the time-evolution operator U is unitary.

Exercise 2.9 Show that any unitary operator U can be written as $U = \exp(iA)$, where A is an Hermitian operator.

Postulate II: *We associate with any observable A a self-adjoint operator A on the Hilbert space $\mathcal{H}_S$. The only possible outcome of a measurement of the observable A is one of the eigenvalues of the operator A. If we write the eigenvalue equation for the operator A,*

$$A|i\rangle = a_i|i\rangle, \tag{2.124}$$

where $|i\rangle$ is an orthonormal basis of eigenvectors of the operator A, and we expand the state vector $|\psi(t)\rangle$ over this basis:

$$|\psi(t)\rangle = \sum_i c_i(t)|i\rangle, \tag{2.125}$$

then the probability that a measurement of the observable A at time t results in outcome a_i is given by

$$p_i(t) = p(a=a_i \mid t) = \left|\langle i|\psi(t)\rangle\right|^2 = \left|c_i(t)\right|^2. \tag{2.126}$$

Note that, for the sake of simplicity, we have stated Postulate II for the case in which the eigenvalues of A are non-degenerate. We shall consider the case of spectral degeneracies later, before stating Postulate III.

Comments

(i) It is important to note that observables are the quantum analogue of dynamical variables in classical mechanics, such as position, linear and angular momentum and so on. In contrast, other characteristics of a system, such as mass or electric charge, are not in the class of observables, but enter as parameters in the Hamiltonian of the system.

(ii) The following argument should help grasp the reason for which self-adjoint operators are associated with physical observables: the eigenvalues of a self-adjoint operator are real (just as the possible outcomes of a measurement) and its eigenvectors form a complete orthonormal set in the Hilbert space $\mathcal{H}_S$ associated with the system. Since $|\psi(t)\rangle$ has unit norm, we have

$$\sum_i p_i(t) = \sum_i |c_i(t)|^2 = 1, \tag{2.127}$$

and therefore the probabilities are normalized, that is, the total probability of obtaining an outcome from the measurement of the observable

A is equal to 1. It is exactly for this reason that Postulate I requires $|\psi(t)\rangle$ to have unit norm.

(iii) In the particular case in which the state vector $|\psi(t_0)\rangle$ at a given time t_0 coincides with an eigenvector of the operator A with eigenvalue a_i,

$$|\psi(t_0)\rangle = |i\rangle, \qquad (2.128)$$

then a measurement of the observable A at time t_0 gives, with unit probability, outcome a_i. For this reason the eigenvectors of the operator A are also called the eigenstates of A.

(iv) Let us assume that $|\psi_1\rangle$ and $|\psi_2\rangle$ are two distinct, normalized eigenvectors of the operator A, with eigenvalues a_1 and a_2, respectively. The superposition principle tells us that the state

$$|\psi\rangle = \lambda_1|\psi_1\rangle + \lambda_2|\psi_2\rangle, \qquad (2.129)$$

with λ_1 and λ_2 complex numbers, is also an allowed state of the system, provided that $|\lambda_1|^2 + |\lambda_2|^2 = 1$, so that $|\psi\rangle$ has unit norm. Therefore, if the system is described by the state vector $|\psi\rangle$ and we perform a measurement of the observable A, we obtain outcome a_1 with probability $|\lambda_1|^2$ and outcome a_2 with probability $|\lambda_2|^2$. However, we stress that the superposition state $|\psi\rangle$ *is not equivalent* to a *naïve* statistical mixture of the states $|\psi_1\rangle$ and $|\psi_2\rangle$, taken with probabilities $|\lambda_1|^2$ and $|\lambda_2|^2$, respectively. We only say that we have a statistical mixture of the states $\{|\psi_i\rangle\}$ (in this case, $|\psi_1\rangle$ and $|\psi_2\rangle$) with weights $\{p_i\}$ (here we have $p_1 = |\lambda_1|^2$ and $p_2 = |\lambda_2|^2$) if the system is in a state taken from the ensemble $\{|\psi_i\rangle\}$, with probabilities $\{p_i\}$. The probabilities $\{p_i\}$ must satisfy the normalization condition $\sum_i p_i = 1$. We shall discuss in detail statistical mixtures in Sec. 5.1. In the following we shall show that a large number N of systems, all in the same state $|\psi\rangle$, is not equivalent to an ensemble of $|\lambda_1|^2 N$ systems in the state $|\psi_1\rangle$ and $|\lambda_2|^2 N$ systems in the state $|\psi_2\rangle$. Indeed, let us assume that we wish to compute the probability $p(b_i)$ of obtaining outcome b_i for some observable B, given that the system is described by the state $|\psi\rangle$. According to Postulate II, we have

$$p(b_i) = \left|\langle i|\psi\rangle\right|^2, \qquad (2.130)$$

where $|i\rangle$ is an eigenvector of the operator B (associated with the observable B) with eigenvalue b_i. Thus, we obtain

$$
\begin{aligned}
p(b_i) &= \left| \lambda_1 \langle i|\psi_1 \rangle + \lambda_2 \langle i|\psi_2 \rangle \right|^2 \\
&= \left| \lambda_1 \right|^2 \left| \langle i|\psi_1 \rangle \right|^2 + \left| \lambda_2 \right|^2 \left| \langle i|\psi_2 \rangle \right|^2 + 2 \operatorname{Re} \left\{ \lambda_1 \lambda_2^\star \langle i|\psi_1 \rangle \langle i|\psi_2 \rangle^\star \right\}.
\end{aligned}
\tag{2.131}
$$

A different result is obtained if we consider a statistical mixture of the states $|\psi_1\rangle$ and $|\psi_2\rangle$, taken with probabilities $|\lambda_1|^2$ and $|\lambda_2|^2$. In this case, the probability $P_{\text{mix}}(b_i)$ of obtaining outcome b_i for a measurement of the observable B is given by

$$
p_{\text{mix}}(b_i) = \left| \lambda_1 \right|^2 \left| \langle i|\psi_1 \rangle \right|^2 + \left| \lambda_2 \right|^2 \left| \langle i|\psi_2 \rangle \right|^2,
\tag{2.132}
$$

and therefore

$$
p(b_i) = p_{\text{mix}}(b_i) + 2 \operatorname{Re} \left\{ \lambda_1 \lambda_2^\star \langle i|\psi_1 \rangle \langle i|\psi_2 \rangle^\star \right\}.
\tag{2.133}
$$

The last term in Eq. (2.133) is called an *interference term*. Therefore, the probability of obtaining b_i as the outcome of a measurement of B, and more generally the predictions of the quantum-mechanical theory, depend not only on the moduli $|\lambda_1|$ and $|\lambda_2|$ but also on the relative phase between the complex numbers λ_1 and λ_2, which affects the product $\lambda_1 \lambda_2^\star$. For example, the four states

$$
|\psi_1\rangle = \tfrac{1}{\sqrt{2}} \left(|0\rangle + |1\rangle \right), \qquad |\psi_2\rangle = \tfrac{1}{\sqrt{2}} \left(|0\rangle - |1\rangle \right),
$$

$$
|\psi_3\rangle = \tfrac{1}{\sqrt{2}} \left(|0\rangle + i|1\rangle \right), \qquad |\psi_4\rangle = \tfrac{1}{\sqrt{2}} \left(|0\rangle - i|1\rangle \right)
\tag{2.134}
$$

represent different states of a system, leading to different experimental outcomes. In contrast, a global phase has no physical significance, that is, the state vectors $|\psi\rangle$ and $e^{i\varphi}|\psi\rangle$, with φ real, give the same predictions for the outcome of any experiment.

Conservative systems

When the Hamiltonian H of a system does not depend explicitly on time, we say that the system is conservative. In this case, we know from classical mechanics that the energy E of the system is constant in time; that is, it is a *constant of motion*. In quantum mechanics, the solution to the Schrödinger equation (2.119) can be written easily, once we know the eigenvalues E_n and eigenvectors $|n\rangle$ of the Hamiltonian operator H. Let us consider the eigenvalue equation for H:

$$
H|n\rangle = E_n|n\rangle,
\tag{2.135}
$$

where, for the sake of simplicity, we assume that the spectrum of the operator H is non-degenerate, that is, $E_m \neq E_n$ for $m \neq n$. Since we have assumed that H does not depend on time, the eigenvalues E_n and eigenvectors $|n\rangle$ are also time-independent. The solution $|\psi(t)\rangle$ to the Schrödinger equation (2.119) can be expanded over the basis of the eigenfunctions of the operator H as follows:

$$|\psi(t)\rangle = \sum_n c_n(t) |n\rangle, \qquad (2.136)$$

with

$$c_n(t) = \langle n|\psi(t)\rangle. \qquad (2.137)$$

The solution $|\psi(t)\rangle$ is uniquely determined by the initial condition $|\psi(t_0)\rangle$, where, for example, we may take $t_0 = 0$. The initial condition $|\psi(0)\rangle$ is determined if the coefficients $c_n(0) = \langle n|\psi(0)\rangle$ are fixed. If we insert expansion (2.136) into the Schrödinger equation (2.119), we obtain

$$i\hbar \frac{d}{dt} c_n = E_n c_n, \qquad (2.138)$$

whose solution is

$$c_n(t) = c_n(0) \exp\left(-\frac{i}{\hbar} E_n t\right). \qquad (2.139)$$

Therefore, the state vector $|\psi(t)\rangle$ at time t is given by

$$|\psi(t)\rangle = \sum_n c_n(0) \exp\left(-\frac{i}{\hbar} E_n t\right) |n\rangle. \qquad (2.140)$$

In the special case in which $|\psi(0)\rangle$ coincides with an eigenvector $|n\rangle$ of the Hamiltonian operator H, $|\psi(0)\rangle = |n\rangle$, the solution (2.140) to the Schrödinger equation reduces to

$$|\psi(t)\rangle = \exp\left(-\frac{i}{\hbar} E_n t\right) |n\rangle. \qquad (2.141)$$

Therefore, the state vectors $|\psi(0)\rangle$ and $|\psi(t)\rangle$ only differ by a global phase factor of no physical significance. For this reason the eigenstates of a time-independent Hamiltonian H are called *stationary states*: if a system is described by such a state, its physical properties do not change in time.

We now discuss the effect of the measurement process on the state of the system. Let us assume that the measurement of an observable A

results in outcome a_n, with a_n a non-degenerate eigenvalue of the self-adjoint operator A. If the measurement does not destroy the system and a new measurement of the observable A immediately follows, we again obtain outcome a_n with unit probability. We can explain this experimental result if we admit that the wave function of the system, which immediately before the first measurement was in the state $|\psi\rangle$, immediately after the measurement *collapses* onto the eigenstate $|n\rangle$ of A associated with the eigenvalue a_n. In the case in which there is degeneracy, we can expand the state $|\psi\rangle$ before the measurement as follows:

$$|\psi\rangle = \sum_n \sum_{s=1}^{g_n} c_{n_s} |n_s\rangle, \tag{2.142}$$

where g_n measures the order of degeneracy of the eigenvalue a_n, that is, the dimension of the subspace spanned by the eigenvectors of A with the same eigenvalue a_n. After a measurement giving outcome a_n, the state of the system belongs to this subspace and is given by

$$\frac{1}{\sqrt{\sum_{s=1}^{g_n} |c_{n_s}|^2}} \sum_{s=1}^{g_n} c_{n_s} |n_s\rangle. \tag{2.143}$$

This state is the normalized projection of $|\psi\rangle$ over the subspace corresponding to the eigenvalue a_n (*i.e.*, spanned by the eigenvectors of A with eigenvalue a_n). We may now state the following postulate:

Postulate III: *If a system is described by the wave vector $|\psi\rangle$ and we measure an observable A, obtaining the outcome a_n, then immediately after the measurement the state of the system is given by*

$$\frac{P_n|\psi\rangle}{\sqrt{\langle\psi|P_n|\psi\rangle}}, \tag{2.144}$$

where P_n is the projection operator over the subspace corresponding to a_n.

Note that, if the wave vector $|\psi\rangle$ is given by Eq. (2.142), then the projector P_n reads

$$P_n = \sum_{s=1}^{g_n} |n_s\rangle\langle n_s|. \tag{2.145}$$

Since the eigenvectors of A constitute an orthonormal basis for the Hilbert space $\mathcal{H}_S$ associated with the system, it is easy to check that the projectors

P_n satisfy the completeness relation,

$$\sum_n P_n = I \,, \tag{2.146}$$

and the orthogonality condition

$$P_n P_m = \delta_{mn} P_m \,. \tag{2.147}$$

In the case without degeneracy, $g_n = 1$, the wave function of the system after the measurement collapses onto the state

$$\frac{1}{|c_n|} c_n |n\rangle \,, \tag{2.148}$$

and therefore (neglecting a global phase factor of no physical significance) onto the eigenstate $|n\rangle$ corresponding to the eigenvalue a_n.

If the system is described by the wave vector (2.142), then upon measuring the observable A, the probability of obtaining any given outcome a_n is given by

$$p_n = \langle\psi|P_n|\psi\rangle \,. \tag{2.149}$$

It is easy to check that for p_n we recover the statement of Postulate II, namely, Eq. (2.126), in the non-degenerate case ($g_n = 1$).

Probability theory now tells us that the average value of the observable A is given by

$$\langle A \rangle = \sum_n a_n p_n \,, \tag{2.150}$$

and therefore

$$\langle A \rangle = \sum_n a_n \langle\psi|P_n|\psi\rangle = \langle\psi|\left(\sum_n a_n P_n\right)|\psi\rangle = \langle\psi|A|\psi\rangle \,, \tag{2.151}$$

where we have used the spectral decomposition $A = \sum_n a_n P_n$.

The standard deviation ΔA associated with observations of A is given by

$$\Delta A = \sqrt{\langle(A - \langle A \rangle)^2\rangle} = \sqrt{\langle A^2 \rangle - \langle A \rangle^2} \,. \tag{2.152}$$

Therefore, if we perform a large number of experiments in which the state $|\psi\rangle$ is prepared and the observable A is measured, we obtain outcomes with mean value $\langle A \rangle$ and standard deviation ΔA.

Starting from the analysis of a few ideal experiments, Heisenberg showed that it is not possible to simultaneously assign a well-determined position and velocity to a given particle. If we increase the precision in our measurement of the particle's velocity, then we increase the uncertainty in its position and *vice versa*. This intrinsically quantum limitation is expressed by the position–momentum uncertainty relations of Heisenberg:

$$\Delta x \, \Delta p_x \geq \frac{\hbar}{2}, \qquad \Delta y \, \Delta p_y \geq \frac{\hbar}{2}, \qquad \Delta z \, \Delta p_z \geq \frac{\hbar}{2}, \qquad (2.153)$$

where Δx, Δy, Δz and Δp_x, Δp_y, Δp_z are the uncertainties in the position and the momentum of the particle. In the following we give the precise mathematical formulation of Heisenberg's uncertainty principle, due to Jordan.

The Heisenberg uncertainty principle: *Suppose that A and B are Hermitian operators associated with observables and $|\psi\rangle$ is a given quantum state. Then the following inequality is satisfied:*

$$\Delta A \, \Delta B \geq \frac{\left| \langle \psi | [A, B] | \psi \rangle \right|}{2}. \qquad (2.154)$$

Proof. Let us consider the operators P and Q, defined by $P = A - \langle A \rangle$ and $Q = B - \langle B \rangle$. We can always write the complex number $\langle \psi | PQ | \psi \rangle$ as equal to $a + ib$, with a and b real numbers. Thus, the average values of the commutator $[P, Q]$ and the anti-commutator $\{P, Q\}$ are given by $\langle \psi | [P, Q] | \psi \rangle = 2ib$ and $\langle \psi | \{P, Q\} | \psi \rangle = 2a$. This implies that

$$\left| \langle \psi | [P, Q] | \psi \rangle \right|^2 \leq \left| \langle \psi | [P, Q] | \psi \rangle \right|^2 + \left| \langle \psi | \{P, Q\} | \psi \rangle \right|^2 = 4(a^2 + b^2)$$
$$= 4 \left| \langle \psi | PQ | \psi \rangle \right|^2 \leq 4 \langle \psi | P^2 | \psi \rangle \langle \psi | Q^2 | \psi \rangle, \qquad (2.155)$$

where the last inequality is the Cauchy–Schwartz inequality proved in the previous section. Finally, we consider the first and the last term in Eq. (2.155). Since $\langle [P, Q] \rangle = \langle [A, B] \rangle$, $\langle P^2 \rangle = (\Delta A)^2$ and $\langle Q^2 \rangle = (\Delta B)^2$, we have proved the Heisenberg inequality (2.154). □

The Heisenberg principle tells us that, given two non-commuting observables A and B, there is an intrinsic limit to the accuracy of the simultaneous measurement of both A and B. The measurement of one observable necessarily disturbs the other. For instance, if the system is prepared in an eigenstate of A associated with a well-determined eigenvalue a_i, a measurement of the observable A always results in outcome a_i. However, if we measure B, the system wave vector collapses onto an eigenstate of B, which is no longer an eigenstate of A, if A and B do not commute. Therefore, if we now measure A again, we obtain different outcomes, with probabilities determined by Postulate II. In quantum mechanics the measurement process disturbs the system: if the observable A is measured to some accuracy ΔA, the observable B is disturbed by some amount ΔB and $\Delta A \Delta B$ satisfies the Heisenberg inequality (2.154). Given two non-commuting observables A and B, it is impossible to measure at the same time both A and B to an arbitrary degree of accuracy: increasing the accuracy in A implies that the accuracy in B diminishes and *vice versa*. There is no similar phenomenon in classical mechanics and we shall see in Chap. 4 that this intrinsically quantum-mechanical result finds applications in the field of cryptography.

Exercise 2.10 Assume that the observables σ_x and σ_y are measured when a system is in the state $|0\rangle$, where $|0\rangle$ denotes the eigenstate of σ_z corresponding to the eigenvalue $+1$. Show that the uncertainty principle implies that $\Delta\sigma_x \Delta\sigma_y \geq 1$.

The Stern–Gerlach experiment, described in Sec. 2.1, is an example of state measurement/preparation. If the apparatus is oriented along the z-axis, we obtain one out of two possible states, $|0\rangle$ or $|1\rangle$. These states are the eigenvectors of the Pauli operator σ_z, corresponding to the eigenvalues $+1$ and -1. If we block the state $|1\rangle$, then we are left with the eigenstate $|0\rangle$ of the spin operator σ_z (see Fig. 2.2). If instead the apparatus is oriented along the x-axis, we obtain one out of two possible states, which are the eigenvectors of the Pauli operator σ_x, that is, $|+\rangle_x = \frac{1}{\sqrt{2}}(|0\rangle + |1\rangle)$ and $|-\rangle_x = \frac{1}{\sqrt{2}}(|0\rangle - |1\rangle)$, corresponding to the eigenvalues $+1$ and -1, respectively.

Note that the most general state of the system can be written as

$$|\psi\rangle = \alpha|0\rangle + \beta|1\rangle, \tag{2.156}$$

with $|\alpha|^2 + |\beta|^2 = 1$. If we introduce the spherical polar angles θ and ϕ (see Fig. 2.6), this state can equivalently be written as

$$|\psi\rangle = \cos\frac{\theta}{2} e^{-i\phi/2} |0\rangle + \sin\frac{\theta}{2} e^{i\phi/2} |1\rangle, \tag{2.157}$$

with $0 \leq \theta \leq \pi$ and $0 \leq \phi < 2\pi$. Such a state is obtained, in the Stern–Gerlach experiment, when the apparatus is directed along the axis singled out by the unit vector $\boldsymbol{u} = (\sin\theta\cos\phi, \sin\theta\sin\phi, \cos\theta)$. Indeed, the state

$|\psi\rangle$ is an eigenstate of the operator

$$\sigma_{\boldsymbol{u}} = \boldsymbol{\sigma} \cdot \boldsymbol{u} = \sigma_x \sin\theta \cos\phi + \sigma_y \sin\theta \sin\phi + \sigma_z \cos\theta \,, \qquad (2.158)$$

where $\boldsymbol{\sigma} = (\sigma_x, \sigma_y, \sigma_z)$. The operator $\sigma_{\boldsymbol{u}}$ has the following matrix representation in the basis of the eigenvectors of σ_z:

$$\sigma_{\boldsymbol{u}} = \begin{pmatrix} \cos\theta & \sin\theta\,e^{-i\phi} \\ \sin\theta\,e^{i\phi} & -\cos\theta \end{pmatrix}. \qquad (2.159)$$

It is easy to check that the matrix σ_u has eigenvectors

$$\begin{aligned} |+\rangle_{\boldsymbol{u}} &= \cos\tfrac{\theta}{2}\,e^{-i\phi/2}\,|0\rangle + \sin\tfrac{\theta}{2}\,e^{i\phi/2}\,|1\rangle\,, \\ |-\rangle_{\boldsymbol{u}} &= -\sin\tfrac{\theta}{2}\,e^{-i\phi/2}\,|0\rangle + \cos\tfrac{\theta}{2}\,e^{i\phi/2}\,|1\rangle\,, \end{aligned} \qquad (2.160)$$

corresponding to the eigenvalues $+1$ and -1, respectively.

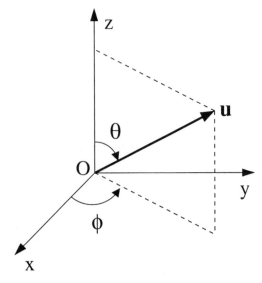

Fig. 2.6   Definition of the spherical polar coordinates θ and ϕ characterizing a unit vector $\boldsymbol{u}$.

The Stern–Gerlach apparatus can be used both to prepare and to measure a state. In the first case we say that the Stern–Gerlach apparatus is used as a *polarizer*, in the latter as an *analyzer*. Let us assume that a beam of atoms of spin-$\frac{1}{2}$ enters a Stern–Gerlach apparatus oriented along the x axis. As we saw in Sec. 2.1, the two components $|+\rangle_x$ and $|-\rangle_x$ come out of the apparatus. If we block component $|-\rangle_x$, then we can say that we have prepared the state $|+\rangle_x$ and in this case the apparatus has been used as a polarizer. If a beam of atoms enters the apparatus oriented, for example, along z, we measure the value of σ_z and the apparatus acts as an analyzer. If the incoming state is described, for instance, by $|\psi\rangle = |+\rangle_x = \frac{1}{\sqrt{2}}(|0\rangle + |1\rangle)$, then the system is in a superposition of the two eigenstates $|0\rangle$ and $|1\rangle$ of σ_z and we can easily predict from Postulate II that the measurement of σ_z will give the eigenvalues $+1$ or -1 of σ_z with equal probabilities $p_+ = p_- = \frac{1}{2}$. Indeed, we have

$$p_+ = |\langle 0|\psi\rangle|^2 = \langle \psi|P_0|\psi\rangle = \tfrac{1}{2},$$

$$p_- = |\langle 1|\psi\rangle|^2 = \langle \psi|P_1|\psi\rangle = \tfrac{1}{2}, \qquad (2.161)$$

where

$$P_0 = |0\rangle\langle 0| = \begin{bmatrix} 1 \\ 0 \end{bmatrix} \begin{bmatrix} 1 & 0 \end{bmatrix} = \begin{bmatrix} 1 & 0 \\ 0 & 0 \end{bmatrix},$$

$$P_1 = |1\rangle\langle 1| = \begin{bmatrix} 0 \\ 1 \end{bmatrix} \begin{bmatrix} 0 & 1 \end{bmatrix} = \begin{bmatrix} 0 & 0 \\ 0 & 1 \end{bmatrix} \qquad (2.162)$$

are the projection operators onto the subspaces spanned by the vectors $|0\rangle$ and $|1\rangle$. It is easy to check that

$$P_0 + P_1 = I = \begin{bmatrix} 1 & 0 \\ 0 & 1 \end{bmatrix}, \qquad P_0 P_1 = 0 = \begin{bmatrix} 0 & 0 \\ 0 & 0 \end{bmatrix}, \qquad (2.163)$$

and therefore the projectors P_0 and P_1 satisfy both the completeness relation (2.146) and the orthogonality condition (2.147).

Exercise 2.11 Show that the results of the Stern–Gerlach experiment illustrated in Fig. 2.4 are in agreement with the predictions of quantum mechanics.

Exercise 2.12 The Schrödinger equation describing the time evolution of the wave vector associated with a spin-half particle of magnetic moment μ in a magnetic field $\boldsymbol{H} = (H_x, H_y, H_z)$ is given by

$$i\hbar\frac{d}{dt}\begin{bmatrix} a(t) \\ b(t) \end{bmatrix} = -\mu(H_x\sigma_x + H_y\sigma_y + H_z\sigma_z)\begin{bmatrix} a(t) \\ b(t) \end{bmatrix}. \qquad (2.164)$$

Solve this equation and compute the mean values of the Pauli operators as a function of time. If the initial wave vector is given by the σ_z eigenstate $|0\rangle$, what magnetic field and evolution time are required to evolve it into the other σ_z eigenstate, namely, $|1\rangle$?

2.5   The EPR paradox and Bell's inequalities

The most spectacular and counter-intuitive manifestation of quantum mechanics is the phenomenon of *entanglement*, observed in composite quantum systems. Let us now discuss the problem. The Hilbert space $\mathcal{H}$ associated with a composite system is the tensor product of the Hilbert spaces $\mathcal{H}_i$ associated with the system's components i. In the simplest case of a bipartite quantum system, we have

$$\mathcal{H} = \mathcal{H}_1 \otimes \mathcal{H}_2 \,. \tag{2.165}$$

The most natural basis for the Hilbert space $\mathcal{H}$ is constructed from the tensor products of the basis vectors of $\mathcal{H}_1$ and $\mathcal{H}_2$. If, for example, the Hilbert spaces $\mathcal{H}_1$ and $\mathcal{H}_2$ are two-dimensional and

$$\{|0\rangle_1, |1\rangle_1\} \,, \qquad \{|0\rangle_2, |1\rangle_2\} \tag{2.166}$$

denote their basis vectors, then a basis for the Hilbert space $\mathcal{H}$ is given by the four vectors

$$\{|0\rangle_1 \otimes |0\rangle_2, \ |0\rangle_1 \otimes |1\rangle_2, \ |1\rangle_1 \otimes |0\rangle_2, \ |1\rangle_1 \otimes |1\rangle_2\} \,. \tag{2.167}$$

The superposition principle tells us that the most general state in the Hilbert space $\mathcal{H}$ is not a tensor product of states residing in $\mathcal{H}_1$ and $\mathcal{H}_2$, but an arbitrary superposition of such states, which we can write as follows:

$$|\psi\rangle = \sum_{i,j=0}^{1} c_{ij} |i\rangle_1 \otimes |j\rangle_2 \,. \tag{2.168}$$

In order to simplify notation, we may also write

$$|\psi\rangle = \sum_{i,j} c_{ij} |ij\rangle \,, \tag{2.169}$$

where the first index in $|ij\rangle$ refers to a state residing in the Hilbert space $\mathcal{H}_1$ and the second to a state in $\mathcal{H}_2$. By definition, a state in $\mathcal{H}$ is said to be *entangled*, or *non-separable*, if it cannot be written as a simple tensor product of a state $|\alpha\rangle_1$ belonging to $\mathcal{H}_1$ and a state $|\beta\rangle_2$ belonging to $\mathcal{H}_2$. In contrast, if we can write

$$|\psi\rangle = |\alpha\rangle_1 \otimes |\beta\rangle_2 \,, \tag{2.170}$$

we say that the state $|\psi\rangle$ is *separable*. As simple examples, let us consider the state

$$|\psi_1\rangle = \tfrac{1}{\sqrt{2}}\left(|00\rangle + |11\rangle\right), \tag{2.171}$$

which is entangled, and the state

$$|\psi_2\rangle = \tfrac{1}{\sqrt{2}}\left(|01\rangle + |11\rangle\right), \tag{2.172}$$

which is separable, since we can write

$$|\psi_2\rangle = \tfrac{1}{\sqrt{2}}\left(|0\rangle + |1\rangle\right) \otimes |1\rangle. \tag{2.173}$$

Exercise 2.13   Show that the state (2.171) is entangled.

When two systems are entangled, it is not possible to assign them individual state vectors $|\alpha\rangle_1$ and $|\beta\rangle_2$. The intriguing non-classical properties of entangled states were clearly illustrated by Einstein, Podolsky and Rosen (EPR) in 1935. These authors showed that quantum theory leads to a contradiction, provided that we accept the following two, seemingly natural, assumptions:

(i) *Reality principle*: If we can predict with certainty the value of a physical quantity, then this value has physical reality, independently of our observation. For example, if a system's wave function $|\psi\rangle$ is an eigenstate of an operator A, namely,

$$A|\psi\rangle = a|\psi\rangle, \tag{2.174}$$

then the value a of the observable A is an element of physical reality

(ii) *Locality principle*: If two systems are causally disconnected, the result of any measurement performed on one system cannot influence the result of a measurement performed on the second system. Following the theory of relativity, we say that two measurement events are disconnected if $(\Delta x)^2 > c^2(\Delta t)^2$, where Δx and Δt are the space and time separations of the two events in some inertial reference frame and c is the speed of light (the two events take place at space-time coordinates (x_1, t_1) and (x_2, t_2), respectively, and $\Delta x = x_2 - x_1$, $\Delta t = t_2 - t_1$).

In quantum mechanics, if an operator B does not commute with A, then the two physical quantities corresponding to the operators A and B cannot have simultaneous reality since we cannot predict with certainty the outcome of the simultaneous measurement of both A and B. Following Heisenberg's principle, a measurement of A destroys knowledge of B.

Let us illustrate the EPR paradox by means of the following simple example, first introduced by Bohm. Consider a source S that emits a pair of spin-$\frac{1}{2}$ particles in the entangled state

$$|\psi\rangle = \tfrac{1}{\sqrt{2}}\left(|01\rangle - |10\rangle\right). \tag{2.175}$$

This state is called an EPR or Bell state. We also say that the system is in a spin-singlet state. One spin-$\frac{1}{2}$ particle is sent to an observer called Alice and the second to another observer called Bob (see Fig. 2.7). Note that Alice and Bob may be located arbitrarily far away from each other. The only requirement is that the measurements performed by Alice and Bob be causally disconnected.

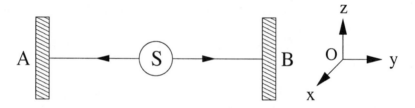

Fig. 2.7   Schematic drawing of the EPR *gedanken* experiment.

If Alice measures the z component of the spin of the particle in her possession and obtains, for instance, $\sigma_z^{(A)} = +1$, then the EPR state collapses onto the state $|01\rangle$ (we remind the reader that the states $|0\rangle$ and $|1\rangle$ are eigenstates of σ_z, corresponding to the eigenvalues $+1$ and -1, respectively). Subsequently, if Bob measures the z component of the spin for his particle, he will obtain $\sigma_z^{(B)} = -1$ with unit probability. Therefore, the results of the measurements of Alice and Bob are perfectly anticorrelated. This result is not surprising according to our intuition, since it is easy to find analogous classical situations. As an example, let us consider two balls, one black and the other white. One ball is sent to Alice and the other to Bob. If Alice finds that her ball is black, then Bob will find with certainty that his ball is white. The surprising point comes from the observation that the spin-singlet state (2.175) can be also written as

$$|\psi\rangle = \tfrac{1}{\sqrt{2}}\left(|+-\rangle - |-+\rangle\right), \tag{2.176}$$

where $|+\rangle = \frac{1}{\sqrt{2}}(|0\rangle+|1\rangle)$ and $|-\rangle = \frac{1}{\sqrt{2}}(|0\rangle-|1\rangle)$ are eigenstates of σ_x with eigenvalues $+1$ and -1, respectively. If Alice measures $\sigma_x^{(A)}$ and obtains, for example, the outcome $\sigma_x^{(A)} = +1$, then the EPR state collapses onto $|+-\rangle$ and Bob will obtain with certainty from the measurement of $\sigma_x^{(B)}$ the outcome $\sigma_x^{(B)} = -1$. Therefore, the state of one particle depends upon the nature of the observable measured on the other particle. If Alice measures $\sigma_z^{(A)}$, then the state of Bob's particle collapses onto an eigenstate of $\sigma_z^{(B)}$. In contrast, if Alice measures $\sigma_x^{(A)}$, then the state of Bob's particle collapses onto an eigenstate of $\sigma_x^{(B)}$. Using the EPR language, we say that in the first case we associate an element of physical reality with $\sigma_z^{(B)}$, in the latter with $\sigma_x^{(B)}$. It is impossible to assign simultaneous physical reality to both observables since they do not commute, $[\sigma_x^{(B)}, \sigma_z^{(B)}] \neq 0$.

The main point is that Alice can choose which observable to measure even after the particles have separated. Therefore, according to the locality principle, any measurement performed by Alice cannot modify the state of Bob's particle. Thus, quantum theory leads to a contradiction if we accept the principles both of realism and locality described above. The EPR conclusion was that quantum mechanics is an incomplete theory. It was later proposed that quantum theory be completed by introducing so-called *hidden variables*. The suggestion was that measurement is in reality a deterministic process, which merely appears probabilistic since some degrees of freedom (hidden variables) are not precisely known.

We point out that the standard interpretation of quantum mechanics does not accept Einstein's local realism. The wave function is not seen as a physical object, but just a mathematical tool, useful to predict probabilities for the outcome of experiments.

Exercise 2.14 Prove that the spin-singlet state (2.175) is rotationally-invariant, that is, that it takes the same form

$$|\psi\rangle = \frac{1}{\sqrt{2}}\left(|+\rangle_u|-\rangle_u - |-\rangle_u|+\rangle_u\right) \tag{2.177}$$

for any direction u, the states $|+\rangle_u$ and $|-\rangle_u$ being eigenstates of $\sigma \cdot u$.

This result is actually rather obvious *a priori*: a spin-singlet state corresponds to zero total spin; thus, no spin vector and no preferred direction can be associated with such a state.

Exercise 2.15 Consider the composite system of a pair of spin-$\frac{1}{2}$ particles, described by the wave function

$$|\psi\rangle = \alpha|00\rangle + \beta|01\rangle + \gamma|10\rangle + \delta|11\rangle \quad (|\alpha|^2 + |\beta|^2 + |\gamma|^2 + |\delta|^2 = 1). \tag{2.178}$$

Assume that the spin polarization σ_z (or σ_x) is measured for the first particle. Discuss the effect of the measurement on the system's wave function.

The debate on the physical reality of quantum systems became the subject of experimental investigation after the formulation, in 1964, of Bell's inequalities. These inequalities are obtained assuming the principles of realism and locality. Since it is possible to devise situations in which quantum mechanics predicts a violation of these inequalities, any experimental observation of such a violation excludes the possibility of a local and realistic description of natural phenomena. In short, Bell showed that the principles of realism and locality lead to experimentally testable inequality relations in disagreement with the predictions of quantum mechanics.

It is instructive to derive Bell's inequalities following a simple model proposed by Wigner. Here we follow the presentation of Sakurai (1994). We assume that a source emits a large number of spin pairs in the singlet state (2.175). Alice and Bob each receive a member of each pair and can measure its polarization along any of three axes a, b and c. We divide the particles in groups as follows. If Alice obtains, for instance, outcome $+1$ when she measures $\sigma_a^{(A)}$, $+1$ when she measures $\sigma_b^{(A)}$ and -1 when she measures $\sigma_c^{(A)}$, then we say that the particle belongs to group $(a+, b+, c-)$. We should stress that we are not saying that Alice measures $\sigma_a^{(A)}$, $\sigma_b^{(A)}$ and $\sigma_c^{(A)}$ simultaneously. She may only measure any one of the spin components. For instance, if she measures $\sigma_a^{(A)}$, then she measures neither $\sigma_b^{(A)}$ nor $\sigma_c^{(A)}$. However, according to the reality principle, we may assign well-defined values to the spin components along the three axes, that is, we assume that these values have physical reality *independently of our observation*. Now remember that the results of Alice's and Bob's measurements must be perfectly anticorrelated for a spin-singlet state. Thus, if Alice's particle belongs to group $(a+, b+, c-)$, then Bob's particle has to be in group $(a-, b-, c+)$. The eight mutually exclusive possibilities are shown in Table 2.1.

Let $p(a+, b+)$ denote the probability that Alice obtains $\sigma_a^{(A)} = +1$ and Bob obtains $\sigma_b^{(B)} = +1$. It is clearly seen from Table 2.1 that

$$p(a+, b+) = \frac{N_3 + N_4}{N_t}, \qquad (2.179)$$

where $N_t \equiv \sum_{i=1}^{8} N_i$. Similarly, we obtain

$$p(a+, c+) = \frac{N_2 + N_4}{N_t}, \qquad p(c+, b+) = \frac{N_3 + N_7}{N_t}. \qquad (2.180)$$

Table 2.1 Division of the spin-singlet states into mutually exclusive groups.

Population	Alice's particle	Bob's particle
N_1	$(a+, b+, c+)$	$(a-, b-, c-)$
N_2	$(a+, b+, c-)$	$(a-, b-, c+)$
N_3	$(a+, b-, c+)$	$(a-, b+, c-)$
N_4	$(a+, b-, c-)$	$(a-, b+, c+)$
N_5	$(a-, b+, c+)$	$(a+, b-, c-)$
N_6	$(a-, b+, c-)$	$(a+, b-, c+)$
N_7	$(a-, b-, c+)$	$(a+, b+, c-)$
N_8	$(a-, b-, c-)$	$(a+, b+, c+)$

Since $N_i \geq 0$, we have $N_3 + N_4 \leq (N_2 + N_4) + (N_3 + N_7)$ and therefore we obtain the following Bell inequality:

$$p(a+, b+) \leq p(a+, c+) + p(c+, b+). \tag{2.181}$$

We point out that we have assumed the locality principle to derive this inequality. Indeed, if a pair belongs to group 1 and Alice chooses to measure $\sigma_a^{(A)}$, then she will certainly obtain outcome 1, independently of the fact that Bob might choose to perform a measurement along the axes a, b or c.

We now evaluate the probabilities appearing in Bell's inequality (2.181) following quantum theory. Let us consider $p(a+, b+)$. If Alice finds $\sigma_a^{(A)} = +1$, then the state of Bob's particle collapses onto the eigenstate $|-\rangle_a$ of $\sigma_a^{(B)}$ with eigenvalue -1. Thus, provided that $\sigma_a^{(A)} = +1$, it is easy to check that Bob obtains $\sigma_b^{(B)} = +1$ with probability $\left| {}_b\langle +|-\rangle_a \right|^2 = \sin^2(\theta_{ab}/2)$, where θ_{ab} is the angle between the axes a and b. Since Alice obtains $\sigma_a^{(A)} = +1$ with probability one half, we obtain

$$p(a+, b+) = \frac{1}{2} \sin^2 \left(\frac{\theta_{ab}}{2} \right). \tag{2.182}$$

In the same way we can compute $p(a+, c+)$ and $p(c+, b+)$. Hence, Bell's inequality (2.181) gives

$$\sin^2 \left(\frac{\theta_{ab}}{2} \right) \leq \sin^2 \left(\frac{\theta_{ac}}{2} \right) + \sin^2 \left(\frac{\theta_{cb}}{2} \right). \tag{2.183}$$

If we choose the axes a, b and c such that $\theta_{ab} = 2\theta$, $\theta_{ac} = \theta_{cb} = \theta$, then this inequality is violated for $0 < \theta < \frac{\pi}{2}$. Therefore, quantum mechanics leads to an experimentally testable violation of Bell's inequalities.

We now give an alternative derivation of Bell's inequalities. Let us assume that there exists a hidden variable λ such that, for any value of λ, a well-defined (deterministic) result $O(\lambda)$ is obtained from the measurement of a physical observable O. We require that the distribution probability $\rho(\lambda)$ of the variable λ be such that the average values predicted by quantum mechanics are recovered; that is,

$$\langle O \rangle = \int O(\lambda)\, \rho(\lambda)\, d\lambda. \tag{2.184}$$

Let us consider the EPR *gedanken* experiment drawn schematically in Fig. 2.7. We call $A(\boldsymbol{a}, \lambda)$ and $B(\boldsymbol{b}, \lambda)$ the results of the measurements of the (causally disconnected) spin polarizations $\boldsymbol{\sigma}^{(A)} \cdot \boldsymbol{a}$ and $\boldsymbol{\sigma}^{(B)} \cdot \boldsymbol{b}$ along the directions $\boldsymbol{a}$ and $\boldsymbol{b}$, performed by Alice and Bob, respectively. Assuming the locality principle, the outcome of Alice's measurements cannot depend on the outcome of Bob's measurements. Therefore, the mean value of the correlations between their polarization measurements is given by

$$C(\boldsymbol{a}, \boldsymbol{b}) = \int A(\boldsymbol{a}, \lambda) B(\boldsymbol{b}, \lambda)\, \rho(\lambda)\, d\lambda. \tag{2.185}$$

For example, as we have seen above, quantum mechanics predicts perfect anticorrelation for the EPR state (2.175) when $\boldsymbol{a} = \boldsymbol{b}$ and therefore

$$C(\boldsymbol{a}, \boldsymbol{a})_{\text{quantum}} = -1. \tag{2.186}$$

Let us compute

$$
\begin{aligned}
C(\boldsymbol{a}, \boldsymbol{b}) - C(\boldsymbol{a}, \boldsymbol{b}') &= \int \big[A(\boldsymbol{a}, \lambda) B(\boldsymbol{b}, \lambda) - A(\boldsymbol{a}, \lambda) B(\boldsymbol{b}', \lambda) \big] \rho(\lambda)\, d\lambda \\
&= \int A(\boldsymbol{a}, \lambda) B(\boldsymbol{b}, \lambda) \big[1 \pm A(\boldsymbol{a}', \lambda) B(\boldsymbol{b}', \lambda) \big] \rho(\lambda)\, d\lambda \\
&\quad - \int A(\boldsymbol{a}, \lambda) B(\boldsymbol{b}', \lambda) \big[1 \pm A(\boldsymbol{a}', \lambda) B(\boldsymbol{b}, \lambda) \big] \rho(\lambda)\, d\lambda.
\end{aligned}
\tag{2.187}
$$

Since $A(\boldsymbol{a}, \lambda)$ and $B(\boldsymbol{b}, \lambda)$ are polarization measurements, we have

$$\big| A(\boldsymbol{a}, \lambda) \big| = 1, \qquad \big| B(\boldsymbol{b}, \lambda) \big| = 1. \tag{2.188}$$

Moreover, $\rho(\lambda)$ is a probability distribution and therefore is non-negative for any λ. Thus, we have

$$\left|C(a,b) - C(a,b')\right| \le \int \left[1 \pm A(a',\lambda)B(b',\lambda)\right]\rho(\lambda)\,d\lambda$$
$$+ \int \left[1 \pm A(a',\lambda)B(b,\lambda)\right]\rho(\lambda)\,d\lambda. \qquad (2.189)$$

This implies that

$$\left|C(a,b) - C(a,b')\right| \le \pm\left[C(a',b') + C(a',b)\right] + 2\int \rho(\lambda)\,d\lambda \qquad (2.190)$$

and therefore

$$\left|C(a,b) - C(a,b')\right| \le -\left|C(a',b') + C(a',b)\right| + 2\int \rho(\lambda)\,d\lambda. \qquad (2.191)$$

We finally obtain

$$\left|C(a,b) - C(a,b')\right| + \left|C(a',b) + C(a',b')\right| \le 2, \qquad (2.192)$$

where we have used the normalization of the probability distribution $\rho(\lambda)$, that is, $\int \rho(\lambda)\,d\lambda = 1$. Inequality (2.192) is known as the CHSH inequality, after its four discovers (Clauser, Horne, Shimony and Holt). It is an example of a larger set known as Bell's inequalities. The main point is that there exist directions (a, b, a', b') such that, considering entangled states, quantum mechanics violates the CHSH inequality. For instance, we may consider the set of directions a, a', b, b' shown in Fig. 2.8. For the spin-singlet state (2.175), quantum mechanics predicts that $C(a,b) = -a \cdot b = -\cos(\theta_{ab})$, where θ_{ab} is the angle between the directions a and b (see Ex. 2.16 below), thus we have

$$\left\{\left|C(a,b) - C(a,b')\right| + \left|C(a',b) + C(a',b')\right|\right\}_{\text{quantum}}$$
$$= \left|-\cos(\phi) + \cos(3\phi)\right| + \left|-\cos(\phi) - \cos(\phi)\right|$$
$$= 2\sqrt{2} \ge 2, \qquad (2.193)$$

when $\phi = \frac{\pi}{4}$.

Exercise 2.16 Show that the quantum-mechanical mean value of the correlation

$$C(a,b)_{\text{quantum}} = \langle\psi|(\sigma^{(A)} \cdot a)(\sigma^{(B)} \cdot b)|\psi\rangle \qquad (2.194)$$

is equal to $-\boldsymbol{a} \cdot \boldsymbol{b}$ when $|\psi\rangle$ is the EPR state (2.175).

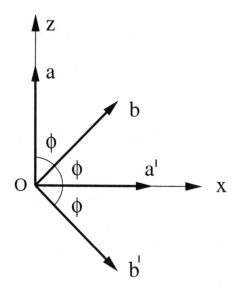

Fig. 2.8   Choice of directions leading to a violation of the CHSH inequality (2.192). The angles labelled ϕ are taken equal to $\frac{\pi}{4}$.

Bell's inequalities represent, first of all, an experimental test of the consistency of quantum mechanics. Many experiments have been performed in order to check Bell's inequalities; the most famous involved EPR pairs of photons and was performed by Aspect and co-workers in 1982. This experiment displayed an unambiguous violation of the CHSH inequality by tens of standard deviations and an excellent agreement with quantum mechanics. More recently, other experiments have come closer to the requirements of the ideal EPR scheme and again impressive agreement with the predictions of quantum mechanics has always been found. Nonetheless, there is no general consensus as to whether or not these experiments may be considered conclusive, owing to the limited efficiency of detectors. If, for the sake of argument, we assume that the present results will not be contradicted by future experiments with high-efficiency detectors, we must conclude that Nature does not experimentally support the EPR point of view. In summary, the World is not locally realistic.

We should stress that there is more to learn from Bell's inequalities and Aspect's experiments than merely a consistency test of quantum mechanics. These profound results show us that entanglement is a fundamentally new resource, beyond the realm of classical physics, and that it is possible to experimentally manipulate entangled states. As we shall see in the following chapters, a major goal of quantum-information science is to exploit this resource to perform computation and communication tasks beyond classical capabilities.

Exercise 2.17 Show that, for a state

$$|\psi\rangle = \alpha|00\rangle + \beta|11\rangle \qquad (|\alpha|^2 + |\beta|^2 = 1), \qquad (2.195)$$

with both α and β different from zero, it is possible to choose the directions of a, a', b and b' so that the CHSH inequality (2.192) is violated. Therefore, violation of Bell's inequalities is a typical feature of entangled states

2.6   A guide to the bibliography

There are many good books on quantum mechanics. An introductory level text is, for instance, Merzbacher (1997) whereas more advanced texts are Sakurai (1994) and the two volumes of Cohen-Tannoudji *et al.* (1977). These books focus on atomic physics topics not directly related to quantum computation and information. A quantum mechanics text closer to quantum information is Peres (1993).

There are many good undergraduate level books on linear algebra. A useful reference is Lang (1996).

The EPR paradox is due to Einstein *et al.* (1935), see also the comment by Bohr (1935). Bell's inequalities were first introduced by Bell (1964). We have presented a particular Bell inequality, known as the CHSH inequality after its four discovers (Clauser *et al.*, 1969). An unambiguous violation of the CHSH inequality was obtained in the experiments by Aspect *et al.* (1981). More recent experiments have come remarkably close to the ideal EPR thought experiment, see, e.g., Weihs *et al.* (1998).

Chapter 3

Quantum Computation

This chapter introduces the basic principles of quantum computation. We shall adhere to the quantum circuit model of computation, with which it is easy to work and which is close to physical implementations. We shall not discuss the more abstract quantum Turing machine model, which, however, has been shown to be equivalent to the circuit model.

The elementary unit of quantum information and the basic building block of quantum computation is the qubit, a two-level quantum system that can be prepared, manipulated and measured in a controlled way. A quantum computer can be seen as a collection of n qubits and therefore its wave function resides in a 2^n-dimensional complex Hilbert space. As far as coupling to the environment may be neglected, its evolution in time is unitary and governed by the Schrödinger equation.

A quantum computation is composed of three basic steps: preparation of the input state, implementation of the desired unitary transformation acting on this state and measurement of the output state. The output of the measurement process is inherently probabilistic and the probabilities of the different possible outputs are set by the basic postulates of quantum mechanics. Therefore, in a quantum algorithm we must, in general, repeat several times the algorithm to obtain the correct solution of our problem with probability as close to one as desired. In this sense, quantum algorithms are analogous to classical probabilistic algorithms. However, the superposition principle and quantum entanglement open up new possibilities for computation. Quantum computers are potentially more powerful than classical (deterministic or probabilistic) computers due to quantum interference and entanglement.

We show that, analogously to classical computation, there exists a small set of gates that are universal, that is, any unitary transformation can be decomposed into a sequence of these gates. We then discuss the implementation of the basic Boolean functions and arithmetic operations on a quantum computer. After that, we focus on quantum algorithms and the techniques underlying their construction. These algorithms take advantage of the basic properties of quantum mechanics, from the superposition principle to entanglement and interference effects, to solve certain computational problems much more efficiently than a classical computer. This includes basic problems of computer science: from the search of a marked item in an unstructured database (Grover's algorithm) to integer factoring (Shor's algorithm). The latter case provides a striking exponential speedup over the best known classical algorithms. After that, we discuss a third relevant class of quantum algorithms, the simulation of physical systems. Finally, we give a brief overview of the first experimental implementations and of their prospects. A much more detailed discussion of this issue will be postponed to Chap. 8.

3.1   The qubit

A classical bit is a system that can exist in two distinct states, which are used to represent 0 and 1, that is, a single binary digit. The only possible operations (gates) in such a system are the identity $(0 \to 0, 1 \to 1)$ and NOT $(0 \to 1, 1 \to 0)$. In contrast, a quantum bit (*qubit*) is a two-level quantum system, described by a two-dimensional complex Hilbert space. In this space, one may choose a pair of normalized and mutually orthogonal quantum states,

$$|0\rangle \equiv \begin{bmatrix} 1 \\ 0 \end{bmatrix}, \qquad |1\rangle \equiv \begin{bmatrix} 0 \\ 1 \end{bmatrix}, \qquad (3.1)$$

to represent the values 0 and 1 of a classical bit. These two states form a *computational basis*. From the superposition principle, any state of the qubit may be written as

$$|\psi\rangle = \alpha|0\rangle + \beta|1\rangle, \qquad (3.2)$$

where the amplitudes α and β are complex numbers, constrained by the normalization condition

$$|\alpha|^2 + |\beta|^2 = 1. \qquad (3.3)$$

Since state vectors are defined only up to a global phase of no physical significance, one may choose α real and positive (except for the basis state $|1\rangle$, in which $\alpha = 0$, and one may take $\beta = 1$ real). Thus, the generic state of a qubit may be written as

$$
\begin{aligned}
|\psi\rangle &= \cos\tfrac{\theta}{2}\,|0\rangle + e^{i\phi}\sin\tfrac{\theta}{2}\,|1\rangle \\
&= \begin{bmatrix} \cos\tfrac{\theta}{2} \\ e^{i\phi}\sin\tfrac{\theta}{2} \end{bmatrix}
\end{aligned}
\qquad (0 \le \theta \le \pi,\ 0 \le \phi < 2\pi). \qquad (3.4)
$$

Therefore, unlike the classical bit, which can only be set equal to 0 or 1, the qubit resides in a vector space, parametrized by the continuous variables α and β (or θ and ϕ). Thus, a continuum of states is allowed. This contradicts our "classical" way of thinking: according to our intuition, a system with two states can only be in one state or in the other. However, as we have seen in Chap. 2, quantum mechanics is much more interesting and allows infinitely many other possibilities. At this stage, one might be tempted to say that a single qubit could be used to store an infinite amount of information. Actually, we must in general provide infinitely many bits to specify the complex numbers α and β in (3.2). However, there is a catch: to extract this information we must perform a measurement and quantum mechanics tells us that from the measurement of the polarization state σ_n of a qubit along any axis n, we obtain only a single bit of information ($\sigma_n = +1$ or $\sigma_n = -1$). Infinitely many measurements on identically prepared single-qubit states are required to obtain α and β.

A two-level quantum system can be used in practice as a qubit if it is possible to manipulate it as follows:

(i) it can be prepared in some well-defined state, for example the state $|0\rangle$, which we call the *fiducial* state of the qubit;

(ii) any state of the qubit can be transformed into any other state. Such transformations are carried out by means of unitary transformations, as we shall see in the next section;

(iii) the qubit state can be measured in the computational basis $\{|0\rangle, |1\rangle\}$. This means that we can measure the qubit polarization along the z-axis. As we have seen in Sec. 2.4, the Hermitian operator associated with this measurement is the Pauli operator σ_z, which has eigenstates $|0\rangle$ and $|1\rangle$. Thus, if the state of the qubit is described by Eq. (3.4), as a result of the measurement one obtains 0 or 1 (that is, $\sigma_z = +1$ or

$\sigma_z = -1$) with probabilities

$$p_0 = \left| \langle 0 | \psi \rangle \right|^2 = \cos^2 \tfrac{\theta}{2}, \qquad p_1 = \left| \langle 1 | \psi \rangle \right|^2 = \sin^2 \tfrac{\theta}{2}, \qquad (3.5)$$

which have been computed using Postulate II of quantum mechanics (discussed in Sec. 2.4).

It is important to stress that, as we shall discuss in detail in Chap. 8, requirements (i)–(iii) can be fulfilled nowadays in laboratory experiments. Physical implementations of a qubit are provided, for instance, by a nuclear spin in a molecule in nuclear magnetic resonance quantum processors, or by the state of an atom in a cavity ($|0\rangle$ corresponds to the atomic ground state and $|1\rangle$ to the first excited state), or by a Cooper pair tunnelling between two superconducting islands ($|0\rangle$ if the pair is on one island and $|1\rangle$ if it is on the other island). The controlled unitary evolution of the state of a qubit is then implemented by means of magnetic or laser fields and efficient measurement apparatuses have now been developed.

3.1.1 *The Bloch sphere*

The Bloch sphere representation is useful in thinking about qubits since it provides a geometric picture of the qubit and of the transformations that one can operate on the state of a qubit. Owing to the normalization condition (3.3), the qubit's state can be represented by a point on a sphere of unit radius, called the *Bloch sphere*. This sphere can be embedded in a three-dimensional space of Cartesian coordinates ($x = \cos\phi\sin\theta$, $y = \sin\phi\sin\theta$, $z = \cos\theta$). Thus, the state (3.4) can be written as

$$|\psi\rangle = \begin{bmatrix} \sqrt{\frac{1+z}{2}} \\ \frac{x+iy}{\sqrt{2(1+z)}} \end{bmatrix}. \qquad (3.6)$$

By definition, a Bloch vector is a vector whose components (x, y, z) single out a point on the Bloch sphere. Therefore, each Bloch vector must satisfy the normalization condition $x^2 + y^2 + z^2 = 1$. We can also say that the angles θ and ϕ define a Bloch vector, as shown in Fig. 3.1. In the same figure, we also show the sinusoidal projection, in which the Bloch sphere is projected onto a plane.[1] The sinusoidal projection helps in visualizing the

---

[1]In this plane (X, Y) the state of a qubit has coordinates $X = \phi\sin\theta$ and $Y = -\theta + \frac{\pi}{2}$ (here the angle variable ϕ has to be taken in the interval $[-\pi, \pi]$). The sinusoidal projection is an area-preserving transformation, the parallels and the $\phi = 0$ meridian of the Bloch sphere are straight lines, all other meridians are sinusoidal curves.

unitary transformations of the state of a qubit.

Another useful representation of the state (3.4) is obtained by means of the projector $P = |\psi\rangle\langle\psi|$. The matrix representation of the operator P on the basis $\{|0\rangle, |1\rangle\}$ is given by

$$
P = \begin{bmatrix} \cos^2 \frac{\theta}{2} & e^{-i\phi} \sin \frac{\theta}{2} \cos \frac{\theta}{2} \\ e^{i\phi} \sin \frac{\theta}{2} \cos \frac{\theta}{2} & \sin^2 \frac{\theta}{2} \end{bmatrix}
$$

$$
= \frac{1}{2} \begin{bmatrix} 1+z & x-iy \\ x+iy & 1-z \end{bmatrix},
\tag{3.7}
$$

where the matrix element P_{ij} $(i, j = 0, 1)$ is defined as $\langle i|P|j\rangle$.

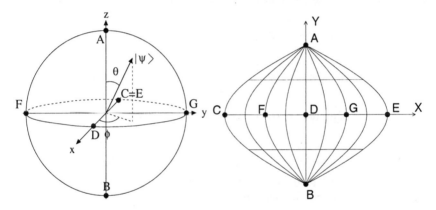

Fig. 3.1 Bloch-sphere representation of a qubit (left) and sinusoidal projection of the Bloch sphere (right). The points corresponding to the following states are shown: $A = (\alpha=1, \beta=0)$, $B = (0, 1)$, $C = E = \left(\frac{1}{\sqrt{2}}, -\frac{1}{\sqrt{2}}\right)$, $D = \left(\frac{1}{\sqrt{2}}, \frac{1}{\sqrt{2}}\right)$, $F = \left(\frac{1}{\sqrt{2}}, -\frac{i}{\sqrt{2}}\right)$, and $G = \left(\frac{1}{\sqrt{2}}, \frac{i}{\sqrt{2}}\right)$. Note that the points A (north pole of the Bloch sphere) and B (south pole) correspond to the states $|0\rangle$ and $|1\rangle$, respectively.

3.1.2 *Measuring the state of a qubit*

The state $|\psi\rangle = \alpha|0\rangle + \beta|1\rangle$ of a qubit can be measured in principle, provided that we have at our disposal a large number of identically prepared qubits. The Bloch-sphere representation offers a particularly well-suited framework to understand this point. In the following, we shall show that the coordinates x, y and z of a qubit on the Bloch sphere can be measured (as we have seen, from these coordinates we can also determine α and β, up to an overall phase factor).

Using the Pauli operators, written in the computational basis as

$$\sigma_x = \begin{bmatrix} 0 & 1 \\ 1 & 0 \end{bmatrix}, \quad \sigma_y = \begin{bmatrix} 0 & -i \\ i & 0 \end{bmatrix}, \quad \sigma_z = \begin{bmatrix} 1 & 0 \\ 0 & -1 \end{bmatrix}, \qquad (3.8)$$

one has, for the state $|\psi\rangle$ given by (3.4),

$$\sigma_x |\psi\rangle = e^{i\phi} \sin \tfrac{\theta}{2} |0\rangle + \cos \tfrac{\theta}{2} |1\rangle ,$$
$$\sigma_y |\psi\rangle = -ie^{i\phi} \sin \tfrac{\theta}{2} |0\rangle + i \cos \tfrac{\theta}{2} |1\rangle , \qquad (3.9)$$
$$\sigma_z |\psi\rangle = \cos \tfrac{\theta}{2} |0\rangle - e^{i\phi} \sin \tfrac{\theta}{2} |1\rangle .$$

Therefore, the following expectation values for the state (3.4) are obtained:

$$\langle\psi|\sigma_x|\psi\rangle = \langle\psi| \begin{bmatrix} 0 & 1 \\ 1 & 0 \end{bmatrix} |\psi\rangle = \sin\theta \cos\phi = x,$$
$$\langle\psi|\sigma_y|\psi\rangle = \langle\psi| \begin{bmatrix} 0 & -i \\ i & 0 \end{bmatrix} |\psi\rangle = \sin\theta \sin\phi = y, \qquad (3.10)$$
$$\langle\psi|\sigma_z|\psi\rangle = \langle\psi| \begin{bmatrix} 1 & 0 \\ 0 & -1 \end{bmatrix} |\psi\rangle = \cos\theta = z.$$

The coordinates (x, y, z) can be obtained with arbitrary accuracy by means of standard projective measurements on the computational basis, that is, measuring σ_z. Indeed, from Eq. (3.5) we obtain

$$p_0 - p_1 = \cos^2 \tfrac{\theta}{2} - \sin^2 \tfrac{\theta}{2} = \cos\theta = z . \qquad (3.11)$$

Thus, the coordinate z is given by the difference of the probabilities to obtain outcomes 0 or 1 from a measurement of σ_z. If we have at our disposal a large number N of systems identically prepared in the state (3.4), we can estimate z as $N_0/N - N_1/N$, where N_0 and N_1 count the number of outcomes 0 and 1. Therefore, z can be measured to any required accuracy, provided we measure a sufficiently large number of states.

The coordinates x and y can be obtained by using the possibility to operate a unitary transformation on the qubit. If the unitary transformation described by the matrix

$$U_1 = \tfrac{1}{\sqrt{2}} \begin{bmatrix} 1 & 1 \\ -1 & 1 \end{bmatrix} \qquad (3.12)$$

is applied to the state (3.4), we obtain the state $|\psi^{(1)}\rangle = U_1|\psi\rangle$. A projective measurement in the computational basis then gives outcome 0 or 1 with probabilities $p_0^{(1)} = |\langle 0|\psi^{(1)}\rangle|^2$ and $p_1^{(1)} = |\langle 1|\psi^{(1)}\rangle|^2$, respectively. Therefore, we obtain

$$p_0^{(1)} - p_1^{(1)} = \cos\phi \sin\theta = x. \tag{3.13}$$

In the same way, if the state (3.4) is transformed by means of the matrix

$$U_2 = \frac{1}{\sqrt{2}} \begin{bmatrix} 1 & -i \\ -i & 1 \end{bmatrix}, \tag{3.14}$$

we obtain the state $|\psi^{(2)}\rangle = U_2|\psi\rangle$. Therefore,

$$p_0^{(2)} - p_1^{(2)} = \sin\phi \sin\theta = y, \tag{3.15}$$

where $p_0^{(2)} = |\langle 0|\psi^{(2)}\rangle|^2$ and $p_1^{(2)} = |\langle 1|\psi^{(2)}\rangle|^2$ give the probabilities to obtain outcome 0 or 1 from the measurement of the qubit polarization along z.

Exercise 3.1 The *fidelity* F of two quantum states $|\psi_1\rangle$ and $|\psi_2\rangle$ is defined by $F \equiv |\langle\psi_1|\psi_2\rangle|^2$. It is a measure of the distance between the two quantum states: We have $0 \leq F \leq 1$, with $F = 1$ when $|\psi_1\rangle$ coincides with $|\psi_2\rangle$ and $F = 0$ when $|\psi_1\rangle$ and $|\psi_2\rangle$ are orthogonal. Show that $F = \cos^2\frac{\alpha}{2}$, with α the angle between the Bloch vectors corresponding to the quantum states $|\psi_1\rangle$ and $|\psi_2\rangle$.

3.2 The circuit model of quantum computation

As was shown in Chap. 1, a classical computer may be described most conveniently as a finite register of n bits. Elementary operations, such as NOT or AND, may be performed on single bits or pairs of bits and these operations may be combined in an ordered way to produce any given complex logic function.

The circuit model can be transferred to quantum computation. The quantum computer may be thought of as a finite collection of n qubits, a *quantum register* of size n. While the state of an n-bit classical computer is described in binary notation by an integer $i \in [0, 2^n - 1]$,

$$i = i_{n-1} 2^{n-1} + \cdots + i_1 2 + i_0, \tag{3.16}$$

with $i_0, i_1, \ldots, i_{n-1} \in [0,1]$ binary digits, the state of an n-qubit quantum computer is

$$|\psi\rangle = \sum_{i=0}^{2^n-1} c_i |i\rangle$$

$$= \sum_{i_{n-1}=0}^{1} \cdots \sum_{i_1=0}^{1} \sum_{i_0=0}^{1} c_{i_{n-1},\ldots,i_1,i_0} |i_{n-1}\rangle \otimes \cdots \otimes |i_1\rangle \otimes |i_0\rangle, \quad (3.17)$$

with the complex numbers c_i constrained by the normalization condition

$$\sum_{i=0}^{2^n-1} |c_i|^2 = 1. \quad (3.18)$$

Therefore, the state of an n-qubit quantum computer is a wave function residing in a 2^n-dimensional Hilbert space, constructed as the tensor product of n 2-dimensional Hilbert spaces, one for each qubit. Taking into account the normalization condition (3.18) and the fact that the state of any quantum system is only defined up to a global phase of no physical significance, the state of the quantum computer is determined by $2(2^n - 1)$ independent real parameters. As an example, we consider the case with $n = 2$ qubits. We write the generic state of a two-qubit quantum computer as

$$\begin{aligned}
|\psi\rangle &= c_0|0\rangle + c_1|1\rangle + c_2|2\rangle + c_3|3\rangle \\
&= c_{0,0}|0\rangle \otimes |0\rangle + c_{0,1}|0\rangle \otimes |1\rangle + c_{1,0}|1\rangle \otimes |0\rangle + c_{1,1}|1\rangle \otimes |1\rangle \\
&= c_{00}|00\rangle + c_{01}|01\rangle + c_{10}|10\rangle + c_{11}|11\rangle, \quad (3.19)
\end{aligned}$$

where, in the last line, we have used the shorthand notation $|i_1 i_0\rangle = |i_1\rangle \otimes |i_0\rangle$. This notation allows us to write the state (3.17) in a simpler way as

$$|\psi\rangle = \sum_{i_{n-1},\ldots,i_1,i_0=0}^{1} c_{i_{n-1}\cdots i_1 i_0} |i_{n-1}\cdots i_1 i_0\rangle. \quad (3.20)$$

The *superposition principle* is clearly visible in Eq. (3.17): while n classical bits can store only a single integer i, the n-qubit quantum register can be prepared in the corresponding state $|i\rangle$ of the computational basis, but also in a superposition. We stress that the number of states of the computational basis in this superposition can be as large as 2^n, which grows exponentially with the number of qubits. The superposition principle opens up new possibilities for computation. When we perform a computation on a classical computer, different inputs require separate runs. In contrast, a quantum computer can perform a computation for exponentially many inputs on a single run. This huge *parallelism* is the basis of the power of quantum computation.

We stress that the superposition principle is not a uniquely quantum feature. Indeed, classical waves satisfying the superposition principle do exist. For instance, we may consider the wave equation for a vibrating string with fixed endpoints. Its solutions $|\varphi_i\rangle$ satisfy the superposition principle and we can write down the most general state $|\varphi\rangle$ of a vibrating string as a linear superposition of these solutions, analogously to Eq. (3.17):

$$|\varphi\rangle = \sum_{i=0}^{2^n-1} c_i |\varphi_i\rangle. \tag{3.21}$$

It is therefore also important to point out the importance of *entanglement* for the power of quantum computation, as compared to any classical computation. Let us discuss the resources necessary to represent the superposition (3.21) in classical versus quantum physics. In order to represent the superposition of 2^n levels classically, these levels must belong to the same system. Indeed, there is no entanglement in classical physics and therefore classical states of separate systems can never be superposed. Thus, we need a number of levels that grows exponentially with n. If Δ is the typical energy separation between two consecutive levels, the amount of energy required for this computation is given by $\Delta 2^n$. Hence, the amount of physical resources needed for the computation grows exponentially with n.[2] In contrast, due to entanglement, in quantum physics a general superposition of 2^n levels may be represented by means of n qubits. Thus, the amount of physical resources (energy) grows only linearly with n.

In order to perform a quantum computation, one should be able to:

(i) *prepare* the quantum computer in a well-defined initial state $|\psi_i\rangle$, which we call the fiducial state of the computer, for instance the state $|0\cdots00\rangle$;

---

[2] Of course, one might imagine classical systems in which the energy levels accumulate below some upper bound. In this case, the amount of energy required for the computation could be considered constant with n. However, we would need measurement devices capable of distinguishing levels that are exponentially close in energy (their typical separation being $\propto 2^{-n}$). It is reasonable to assume that exponentially large physical resources would be required for such a measurement apparatus to work.

(ii) *manipulate* the quantum-computer wave function, that is, drive any given unitary transformation U, leading to $|\psi_f\rangle = U|\psi_i\rangle$;

(iii) perform, at the end of the algorithm, a standard *measurement* in the computational basis, that is, measure the polarization σ_z of each qubit.

Since the quantum computer is an n-body (-qubit) quantum system, the time evolution of the wave function (3.17) is governed by the Schrödinger equation. As a result, the postulates of quantum mechanics discussed in Chap. 2 tell us that the evolution of the quantum-computer wave function is described by a unitary operator. Here we neglect non-unitary decoherence effects due to undesired coupling of the quantum computer to the environment, a problem that we shall consider in Chap. 6.

We emphasize that, even though the evolution of an n-qubit wave function is described by a $2^n \times 2^n$ unitary matrix, this matrix can always be decomposed into a product of unitary operations acting only on one or two qubits. These operations are the elementary *quantum gates* of the circuit model of quantum computation.

Finally, we point out that it is possible to show that any complex collective many-qubit measurement can always be performed in the computational basis, provided that it is preceded by a suitable unitary transformation. An example of such a procedure was given in the previous section for a single qubit: the Bloch-sphere coordinate x (or y) can be obtained if the unitary transformation (3.12), or (3.14), is followed by a projection onto the standard basis $\{|0\rangle, |1\rangle\}$.

3.3   Single-qubit gates

The operations on a qubit must preserve the normalization condition (3.3) and are thus described by 2×2 unitary matrices. In the following, we shall introduce the Hadamard and the phase-shift gates and show that they are sufficient to perform any unitary operation on a single qubit.

The *Hadamard gate* is defined as

$$H = \frac{1}{\sqrt{2}} \begin{bmatrix} 1 & 1 \\ 1 & -1 \end{bmatrix}. \tag{3.22}$$

This gate turns the computational basis $\{|0\rangle, |1\rangle\}$ into the new basis $\{|+\rangle, |-\rangle\}$, whose states are a superposition of the states of the computational basis:

$$H \left| 0 \right\rangle = \tfrac{1}{\sqrt{2}} \left(\left| 0 \right\rangle + \left| 1 \right\rangle \right) \equiv \left| + \right\rangle,$$
$$H \left| 1 \right\rangle = \tfrac{1}{\sqrt{2}} \left(\left| 0 \right\rangle - \left| 1 \right\rangle \right) \equiv \left| - \right\rangle. \tag{3.23}$$

Since $H^2 = I$, the inverse transformation $H^{-1} = H$. Note that H is Hermitian. Indeed, it is evident from the matrix representation (3.22) that $(H^T)^\star = H$.

The *phase-shift gate* is defined as

$$R_z(\delta) = \begin{bmatrix} 1 & 0 \\ 0 & e^{i\delta} \end{bmatrix}. \tag{3.24}$$

This gate turns $\left| 0 \right\rangle$ into $\left| 0 \right\rangle$ and $\left| 1 \right\rangle$ into $e^{i\delta} \left| 1 \right\rangle$. Since global phases have no physical meaning, the states of the computational basis, $\left| 0 \right\rangle$ and $\left| 1 \right\rangle$, are unchanged. However, the action of the phase-shift gate on a generic single-qubit state $\left| \psi \right\rangle$ (Eq. (3.4)), gives

$$R_z(\delta) \left| \psi \right\rangle = \begin{bmatrix} 1 & 0 \\ 0 & e^{i\delta} \end{bmatrix} \begin{bmatrix} \cos \tfrac{\theta}{2} \\ e^{i\phi} \sin \tfrac{\theta}{2} \end{bmatrix} = \begin{bmatrix} \cos \tfrac{\theta}{2} \\ e^{i(\phi+\delta)} \sin \tfrac{\theta}{2} \end{bmatrix}. \tag{3.25}$$

Since, as we have discussed in Sec. 2.4, relative phases are observable, the state of the qubit has been changed by the application of the phase-shift gate. It is easy to recognize from Eq. (3.25) that this gate generates a counterclockwise rotation through an angle δ about the z axis of the Bloch sphere (see Fig. 3.1).

It is important to realize that any unitary operation on a single qubit can be constructed using only Hadamard and phase-shift gates. Actually, a unitary transformation moves the qubit state from one point of the Bloch sphere to another point and this can be obtained using only these two quantum gates. In particular, the generic state (3.4) can be reached starting from $\left| 0 \right\rangle$ as follows:

$$R_z(\tfrac{\pi}{2} + \phi) \, H \, R_z(\theta) \, H \left| 0 \right\rangle = e^{i\frac{\theta}{2}} \left(\cos \tfrac{\theta}{2} \left| 0 \right\rangle + e^{i\phi} \sin \tfrac{\theta}{2} \left| 1 \right\rangle \right). \tag{3.26}$$

Exercise 3.2 Show that the unitary operator moving the state parametrized on the Bloch sphere by the angles (θ_1, ϕ_1) into the state (θ_2, ϕ_2) is given by

$$R_z(\tfrac{\pi}{2} + \phi_2) \, H \, R_z(\theta_2 - \theta_1) \, H \, R_z(-\tfrac{\pi}{2} - \phi_1). \tag{3.27}$$

3.3.1   *Rotations of the Bloch sphere*

We now consider a useful class of unitary transformations, the *rotations* of the Bloch sphere about an arbitrary axis. First of all, we need the following result: Let O be an operator such that $O^2 = I$. Thus, $O^k = I$ for k even and $O^k = O$ for k odd. As a consequence, from a Taylor expansion of the exponential of the operator O, one obtains

$$
\begin{aligned}
e^{-i\alpha O} &= \left[1 - \tfrac{1}{2!}\alpha^2 + \cdots\right] I - i\left[\alpha - \tfrac{1}{3!}\alpha^3 + \cdots\right] O \\
&= \cos(\alpha)\, I - i\sin(\alpha)\, O\,.
\end{aligned} \tag{3.28}
$$

Since the Pauli operators satisfy the condition $\sigma_x^2 = \sigma_y^2 = \sigma_z^2 = I$, we can apply Eq. (3.28) to exponentials of σ_x, σ_y and σ_z. In the case of σ_z, we have

$$
\begin{aligned}
e^{-i\frac{\delta}{2}\sigma_z} &= \cos\tfrac{\delta}{2}\, I - i\sin\tfrac{\delta}{2}\, \sigma_z \\
&= e^{-i\frac{\delta}{2}}
\begin{bmatrix}
1 & 0 \\
0 & e^{i\delta}
\end{bmatrix}
\equiv R_z(\delta)\,.
\end{aligned} \tag{3.29}
$$

We note that the above definition of $R_z(\delta)$ differs from (3.25) only in a global phase factor of no physical significance. If we apply the phase-shift gate to a generic vector $|\psi\rangle$ given by Eq. (3.4), we obtain, as explained in Eq. (3.25), the state

$$
R_z(\delta)\,|\psi\rangle = \cos\tfrac{\theta}{2}\,|0\rangle + e^{i(\phi+\delta)}\,\sin\tfrac{\theta}{2}\,|1\rangle. \tag{3.30}
$$

Thus, if (x, y, z) denote the Cartesian coordinates of the vector $|\psi\rangle$ and (x', y', z') the coordinates of $R_z(\delta)\,|\psi\rangle$ (computed from the Bloch-sphere coordinates as explained in Sec. 3.1), we have the following coordinate transformation:

$$
\begin{cases}
x' = x\cos\delta - y\sin\delta\,, \\
y' = x\sin\delta + y\cos\delta\,, \\
z' = z\,.
\end{cases} \tag{3.31}
$$

Thus, $R_z(\delta)$ corresponds to a counterclockwise rotation through an angle δ about the z-axis of the Bloch sphere. Of course, we could equivalently say that the vector $|\psi\rangle$ is rotated counterclockwise through an angle δ or that the Bloch sphere itself is rotated clockwise through an angle δ. In the first picture, we imagine the motion of the vector on a fixed Bloch sphere, in the latter picture we consider a fixed vector and moving axes. Analogously, one can obtain the unitary matrices corresponding to counterclockwise rotations about the x-axis:

$$e^{-i\frac{\delta}{2}\sigma_x} = \begin{bmatrix} \cos\frac{\delta}{2} & -i\sin\frac{\delta}{2} \\ -i\sin\frac{\delta}{2} & \cos\frac{\delta}{2} \end{bmatrix} \equiv R_x(\delta), \qquad (3.32)$$

or the y-axis:

$$e^{-i\frac{\delta}{2}\sigma_y} = \begin{bmatrix} \cos\frac{\delta}{2} & -\sin\frac{\delta}{2} \\ \sin\frac{\delta}{2} & \cos\frac{\delta}{2} \end{bmatrix} \equiv R_y(\delta). \qquad (3.33)$$

Exercise 3.3 Check from the corresponding transformation of the Bloch-sphere coordinates that $R_x(\delta)$ and $R_y(\delta)$ correspond to counterclockwise rotations through an angle δ about the axes x and y, respectively.

Rotation about a generic axis is obtained using the property that infinitesimal rotations can be composed as vectors. Rotation through an angle $\epsilon \ll 1$ about the axis directed along the unit vector $n = (n_x, n_y, n_z)$ is given by the operator

$$R_n(\epsilon) \approx R_x(n_x\epsilon) R_y(n_y\epsilon) R_z(n_z\epsilon). \qquad (3.34)$$

Since the Taylor expansion of Eq. (3.28) gives, for $\epsilon \ll 1$,

$$R_i(n_i\epsilon) \approx I - i\frac{\epsilon}{2} n_i \sigma_i, \qquad (3.35)$$

we obtain

$$R_n(\epsilon) \approx I - i\frac{\epsilon}{2}(n \cdot \sigma), \qquad (3.36)$$

where $\sigma = (\sigma_x, \sigma_y, \sigma_z)$. A finite rotation through an angle δ is obtained by the composition of k infinitesimal rotations through an angle $\epsilon = \delta/k$:

$$R_n(\delta) = \lim_{k\to\infty} \left[I - i\frac{\delta}{2k}(n \cdot \sigma)\right]^k = \exp\left[-i\frac{\delta}{2}(n \cdot \sigma)\right]. \qquad (3.37)$$

Since $(n \cdot \sigma)^2 = n_x^2\sigma_x^2 + n_y^2\sigma_y^2 + n_z^2\sigma_z^2 = (n_x^2 + n_y^2 + n_z^2)I = I$, then Eq. (3.28) applies and therefore

$$R_n(\delta) = \cos\frac{\delta}{2} I - i\sin\frac{\delta}{2}(n \cdot \sigma). \qquad (3.38)$$

From this equation it is clear that we can see the Hadamard gate as a rotation through an angle $\delta = \pi$ about the axis $\tilde{n} = \left(\frac{1}{\sqrt{2}}, 0, \frac{1}{\sqrt{2}}\right)$. Indeed,

$$H = \frac{1}{\sqrt{2}}(\sigma_z + \sigma_x), \qquad (3.39)$$

which coincides with $R_{\tilde{n}}(\pi)$, up to an overall phase. This transformation rotates the x-axis to z and vice versa.

Exercise 3.4 Show that the matrices U_1 and U_2, introduced in Eqs. (3.12) and (3.14), correspond to $R_y(-\frac{\pi}{2})$ and $R_x(\frac{\pi}{2})$, respectively.

Exercise 3.5 Taking into account that a generic 2×2 unitary matrix U can be seen (up to an overall phase factor) as a rotation of angle δ about some axis of the Bloch sphere, compute $\sqrt{U}$.

Exercise 3.6 Prove that $(\boldsymbol{a} \cdot \boldsymbol{\sigma})(\boldsymbol{b} \cdot \boldsymbol{\sigma}) = (\boldsymbol{a} \cdot \boldsymbol{b}) I + i\boldsymbol{\sigma} \cdot (\boldsymbol{a} \times \boldsymbol{b})$.

3.4 Controlled gates and entanglement generation

Entanglement, which is the most intriguing characteristic of quantum mechanics, appears already with two qubits. Actually, a generic two-qubit state can be written in the computational basis as

$$|\psi\rangle = \alpha |00\rangle + \beta |01\rangle + \gamma |10\rangle + \delta |11\rangle, \qquad (3.40)$$

with α, β, γ and δ complex coefficients. Taking into account the normalization condition, $|\alpha|^2 + |\beta|^2 + |\gamma|^2 + |\delta|^2 = 1$ and the fact that the state is only defined up to an overall phase factor, there remain 6 real degrees of freedom. Therefore, it is not possible, in general, to consider the state (3.40) as a separable state. Indeed, as discussed in Sec. 2.5, a state $|\psi\rangle$ of a bipartite quantum system is said to be separable when it is possible to write $|\psi\rangle = |\psi_1\rangle \otimes |\psi_0\rangle$, with $|\psi_1\rangle$ and $|\psi_0\rangle$ wave functions for the two subsystems. Therefore, a separable two-qubit state has only 4 degrees of freedom since, for instance, we can take for each qubit the two parameters of its Bloch sphere. The situation becomes more complex on increasing the number of qubits. One may say that the complexity of entanglement grows exponentially with the number of qubits: while a separable state of n qubits depends only on $2n$ real parameters, the most general (entangled) state has $2(2^n - 1)$ degrees of freedom.

It is clear that single-qubit gates are unable to generate entanglement in an n-qubit system. Indeed, if we start from a separable state, $|\psi\rangle = |\psi_{n-1}\rangle \otimes |\psi_{n-2}\rangle \otimes \cdots \otimes |\psi_0\rangle$, we can move at will any qubit on its Bloch sphere, obtaining $|\psi'\rangle = |\psi'_{n-1}\rangle \otimes |\psi'_{n-2}\rangle \otimes \cdots \otimes |\psi'_0\rangle$. Here any state of the type $|\psi_i\rangle$ can be transformed by gates acting on the qubit i in whatever superposition of the states $|0\rangle$ and $|1\rangle$, but the n-qubit state is still separable.

To prepare an entangled state one needs inter-qubit *interactions*, that is, a two-qubit gate. The prototypical two-qubit gate that is able to generate entanglement is the controlled-NOT gate. This gate acts on the states of the computational basis, $\{|i_1 i_0\rangle = |00\rangle, |01\rangle, |10\rangle, |11\rangle\}$, as the classical XOR gate: $\mathrm{CNOT}(|x\rangle|y\rangle) = |x\rangle|x \oplus y\rangle$, with $x, y = 0, 1$ and $\oplus$ indicating addition modulo 2. The first qubit in the CNOT gate acts as a *control* and the second as a *target*. The gate flips the state of the target qubit if the control qubit is in the state $|1\rangle$ and does nothing if the control qubit is in the state $|0\rangle$. We note that, as discussed in Sec. 2.3, the basis vectors can be represented as column vectors:

$$|0\rangle = |00\rangle = \begin{bmatrix} 1 \\ 0 \\ 0 \\ 0 \end{bmatrix}, \qquad |1\rangle = |01\rangle = \begin{bmatrix} 0 \\ 1 \\ 0 \\ 0 \end{bmatrix},$$

$$|2\rangle = |10\rangle = \begin{bmatrix} 0 \\ 0 \\ 1 \\ 0 \end{bmatrix}, \qquad |3\rangle = |11\rangle = \begin{bmatrix} 0 \\ 0 \\ 0 \\ 1 \end{bmatrix}, \tag{3.41}$$

where the vector $|i\rangle$ has the component i equal to one and all other components equal to zero. In binary notation, $|i\rangle = |i_1 i_0\rangle$ and $|j\rangle = |j_1 j_0\rangle$. Therefore, we can find a matrix representation of the CNOT gate:

$$\mathrm{CNOT} = \begin{bmatrix} 1 & 0 & 0 & 0 \\ 0 & 1 & 0 & 0 \\ 0 & 0 & 0 & 1 \\ 0 & 0 & 1 & 0 \end{bmatrix}, \tag{3.42}$$

where the components $(\mathrm{CNOT})_{ij}$ of this matrix are given by $(\mathrm{CNOT})_{ij} = \langle i| \mathrm{CNOT} |j\rangle$ (note that $i, j = 0, \ldots, 3$). For example, we have

$$(\mathrm{CNOT})_{23} = \langle 2| \mathrm{CNOT} |3\rangle = \langle 10| \mathrm{CNOT} |11\rangle = 1. \tag{3.43}$$

Of course, the CNOT gate, in contrast to the classical XOR gate, can also be applied to any superposition of the computational basis states. Note that CNOT is self-inverse, since $(\mathrm{CNOT})^2 = I$.

It is easy to see that CNOT can generate entangled states. For example,

$$\mathrm{CNOT}\,(\alpha|0\rangle + \beta|1\rangle)|0\rangle = \alpha|00\rangle + \beta|11\rangle, \tag{3.44}$$

which is non-separable insofar as $\alpha, \beta \neq 0$.

Exercise 3.7   The most general separable state of two qubits can be written, up to an overall phase, as

$$|\psi\rangle = a\left\{|0\rangle + b_1 e^{i\phi_1}|1\rangle\right\} \otimes \left\{|0\rangle + b_0 e^{i\phi_0}|1\rangle\right\}, \qquad (3.45)$$

where a is set by the wave-function normalization. What conditions should the real coefficients b_0, b_1, ϕ_0 and ϕ_1 satisfy in order that CNOT$|\psi\rangle$ be entangled?

It is also possible to define generalized controlled-NOT gates, depending on whether the control qubit is the first or the second qubit and whether the gate acts trivially when the control qubit is set to $|0\rangle$ or $|1\rangle$ (we say that a gate acts trivially if its action reduces to the identity). Correspondingly, we have the following four matrices:

$$A = \begin{bmatrix} 1 & 0 & 0 & 0 \\ 0 & 1 & 0 & 0 \\ 0 & 0 & 0 & 1 \\ 0 & 0 & 1 & 0 \end{bmatrix}, \qquad B = \begin{bmatrix} 0 & 1 & 0 & 0 \\ 1 & 0 & 0 & 0 \\ 0 & 0 & 1 & 0 \\ 0 & 0 & 0 & 1 \end{bmatrix},$$

$$C = \begin{bmatrix} 1 & 0 & 0 & 0 \\ 0 & 0 & 0 & 1 \\ 0 & 0 & 1 & 0 \\ 0 & 1 & 0 & 0 \end{bmatrix}, \qquad D = \begin{bmatrix} 0 & 0 & 1 & 0 \\ 0 & 1 & 0 & 0 \\ 1 & 0 & 0 & 0 \\ 0 & 0 & 0 & 1 \end{bmatrix}. \qquad (3.46)$$

The first of these matrices (A) is the standard CNOT gate (3.42), B flips the second qubit if the first is set to $|0\rangle$, C flips the first qubit if the second is $|1\rangle$, and D flips the first qubit if the second is $|0\rangle$. The circuit representations for the generalized CNOT gates are given in Fig. 3.2. As usual in these graphical representations, each line corresponds to a qubit and any sequence of logic gates must be read from the left (input) to the right (output). From bottom to top, qubits run from the least significant (i_0, according to binary notation (3.16)) to the most significant (i_{n-1}). Here a qubit is said to be more significant than another if its flip gives a larger variation in the integer number coded by the state of the n qubits.

Exercise 3.8   Show that all four generalized CNOT gates can be constructed using only the standard CNOT gate and single-qubit gates. In particular, check that the circuit represented in Fig. 3.3 interchanges control and target qubits.

Exercise 3.9   Show that all $4! = 24$ permutations of the basis states of two qubits can be obtained using the generalized CNOT gates and draw the

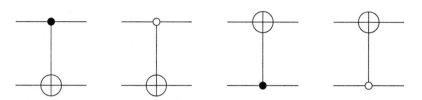

Fig. 3.2 Circuit representations for the generalized CNOT gates. From left to right: A, B, C and D. Note that on the control qubit we draw a full circle if the target qubit is flipped when the control is set to $|1\rangle$, an empty circle if instead the target is flipped when the control is $|0\rangle$.

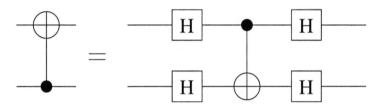

Fig. 3.3 Decomposition of the generalized CNOT gate C into a standard CNOT gate and four Hadamard gates.

corresponding circuits. In particular, check that one can swap two qubits by means of the circuit of Fig. 3.4.

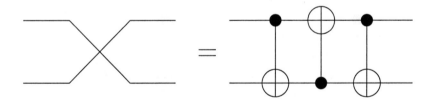

Fig. 3.4 A circuit for the SWAP gate.

Unlike the CNOT, there exist two-qubit quantum gates with no classical analog, for instance the *controlled phase shift*

$$\text{CPHASE}(\delta) = \begin{bmatrix} 1 & 0 & 0 & 0 \\ 0 & 1 & 0 & 0 \\ 0 & 0 & 1 & 0 \\ 0 & 0 & 0 & e^{i\delta} \end{bmatrix}, \tag{3.47}$$

which applies a phase shift to the target qubit only when the control qubit is in the state $|1\rangle$: we have $\text{CPHASE}\,|11\rangle = e^{i\delta}|11\rangle$. We show in Fig. 3.5 that a controlled phase-shift gate can be performed using CNOT gates and single-qubit phase-shift gates.

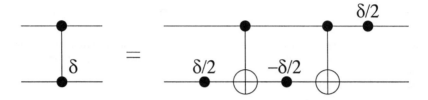

Fig. 3.5   A circuit implementing the controlled phase-shift gate.

Exercise 3.10   The CMINUS gate is defined as $\text{CMINUS} = \text{CPHASE}(\pi)$, that is,

$$\text{CMINUS} = \begin{bmatrix} 1 & 0 & 0 & 0 \\ 0 & 1 & 0 & 0 \\ 0 & 0 & 1 & 0 \\ 0 & 0 & 0 & -1 \end{bmatrix}. \tag{3.48}$$

This gate is important since in some implementations it is easier to perform CMINUS rather than CNOT. Check the relation between CNOT and CMINUS shown in Fig. 3.6

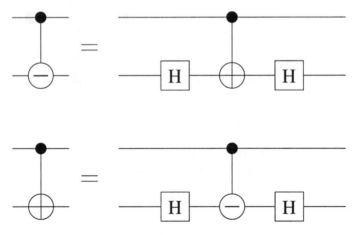

Fig. 3.6   The relation between the CNOT and CMINUS gates. The symbol on the left-hand side of the top circuit denotes CMINUS.

Exercise 3.11 *Backward sign propagation.* We define the amplitude and phase errors as follows. Given an arbitrary state of a qubit, $|\psi\rangle = \alpha|0\rangle + \beta|1\rangle$, the amplitude error performs the transformation

$$|\psi\rangle \rightarrow |\psi_a\rangle = \beta|0\rangle + \alpha|1\rangle, \qquad (3.49)$$

while the effect of the phase error is

$$|\psi\rangle \rightarrow |\psi_p\rangle = \alpha|0\rangle - \beta|1\rangle. \qquad (3.50)$$

Discuss the effect of amplitude and phase errors, acting on the control or target qubit, on the CNOT gate. In particular, consider the initial state

$$(\alpha|0\rangle + \beta|1\rangle) \otimes \tfrac{1}{\sqrt{2}}(|0\rangle + |1\rangle), \qquad (3.51)$$

and show that a phase error acting on the target qubit is also transferred, after application of the CNOT gate, to the control qubit. Note that, as we shall see later in this chapter, this backward sign propagation is also a key ingredient in several quantum algorithms (*e.g.*, Deutsch's and Grover's algorithms).

Exercise 3.12 It is possible to show that the 4×4 Hermitian matrices constitute a linear vector space and that the tensor products $\sigma_i \otimes \sigma_j$ are a basis for this space (where $\sigma_0 \equiv I$, $\sigma_1 \equiv \sigma_x$, $\sigma_2 \equiv \sigma_y$ and $\sigma_3 \equiv \sigma_z$). Therefore, all operators associated with two-qubit observables can be expanded over this basis. Compute the matrix representations of $\sigma_i \otimes \sigma_j$ in the computational basis.

Exercise 3.13 Compute the expectation values of the operators $\sigma_i \otimes \sigma_j$ on the state vector

$$|\psi\rangle = c|00\rangle + \alpha|01\rangle + \beta|10\rangle + \gamma|11\rangle, \qquad (3.52)$$

where the overall phase factor (up to which $|\psi\rangle$ is defined) is chosen so that c is real (while α, β and γ are complex numbers).

3.4.1   *The Bell basis*

As we have shown, the CNOT gate can generate entanglement. In particular, the entangled states of the so-called Bell basis, defined by

$$|\phi^+\rangle = \tfrac{1}{\sqrt{2}}\left(|00\rangle + |11\rangle\right),\tag{3.53a}$$

$$|\phi^-\rangle = \tfrac{1}{\sqrt{2}}\left(|00\rangle - |11\rangle\right),\tag{3.53b}$$

$$|\psi^+\rangle = \tfrac{1}{\sqrt{2}}\left(|01\rangle + |10\rangle\right),\tag{3.53c}$$

$$|\psi^-\rangle = \tfrac{1}{\sqrt{2}}\left(|01\rangle - |10\rangle\right),\tag{3.53d}$$

can be obtained starting from the computational basis, by means of the
circuit represented in Fig. 3.7. It is easy to check that this circuit produces
the following transformations:

$$|00\rangle \to |\phi^+\rangle, \qquad\qquad |01\rangle \to |\psi^+\rangle,$$

$$|10\rangle \to |\phi^-\rangle, \qquad\qquad |11\rangle \to |\psi^-\rangle.\tag{3.54}$$

We note that this transformation can be inverted simply by running the
circuit of Fig. 3.7 from right to left, since both CNOT and Hadamard gates
are self-inverse. As a result, any state of the Bell basis is transformed into a
separable state, this is possible since we have used a two-qubit gate. At this
point it is possible, via a standard measurement in the computational basis,
to establish which of the four Bell states was present at the beginning.

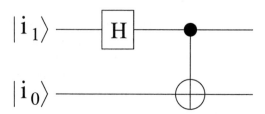

Fig. 3.7   A circuit that transforms the computational basis states $|i_1 i_0\rangle$ to the Bell
states.

3.5   Universal quantum gates

The usefulness of the circuit model in classical computation is due to the
fact that a sequence of elementary operations (*e.g.*, NAND and COPY)
allows one to build up arbitrarily complex computations. In this section,

we shall show that a similar result exists for quantum computation, that is, any unitary operation in the Hilbert space of n qubits can be decomposed into one-qubit and two-qubit CNOT gates. In the following, we shall give a detailed proof of this important result, since it will help the reader to become familiar with quantum logic gate operations.

Let us first define the controlled-U operation. If U is an arbitrary single-qubit unitary transformation, then controlled-U means that U acts on the target qubit only if the control qubit is set to $|1\rangle$:

$$|i_1\rangle|i_0\rangle \rightarrow |i_1\rangle U^{i_1}|i_0\rangle. \qquad (3.55)$$

We now show that the controlled-U gate can be implemented using only single-qubit gates and CNOT. Since a matrix U is unitary if and only if its rows and columns are orthonormal, it turns out that any 2×2 unitary matrix may be written as

$$U = \begin{bmatrix} e^{i(\delta-\alpha/2-\beta/2)}\cos\frac{\theta}{2} & -e^{i(\delta-\alpha/2+\beta/2)}\sin\frac{\theta}{2} \\ e^{i(\delta+\alpha/2-\beta/2)}\sin\frac{\theta}{2} & e^{i(\delta+\alpha/2+\beta/2)}\cos\frac{\theta}{2} \end{bmatrix}, \qquad (3.56)$$

where δ, α, β and θ are real parameters. Therefore, it is possible to decompose U as follows:

$$U = \Phi(\delta)\, R_z(\alpha)\, R_y(\theta)\, R_z(\beta), \qquad (3.57)$$

where

$$\Phi(\delta) = \begin{bmatrix} e^{i\delta} & 0 \\ 0 & e^{i\delta} \end{bmatrix} \qquad (3.58)$$

and R_y and R_z are the rotation matrices about the y and z axes, defined in Eqs. (3.33), (3.29). Indeed, for any U written as in Eq. (3.56), there exist three unitary matrices, A, B and C:

$$A = R_z(\alpha)\, R_y\left(\tfrac{\theta}{2}\right), \qquad (3.59a)$$

$$B = R_y\left(-\tfrac{\theta}{2}\right) R_z\left(-\tfrac{\alpha+\beta}{2}\right), \qquad (3.59b)$$

$$C = R_z\left(\tfrac{\beta-\alpha}{2}\right), \qquad (3.59c)$$

such that

$$ABC = I \quad \text{and} \quad \Phi(\delta)\, A\, \sigma_x\, B\, \sigma_x\, C = U. \qquad (3.60)$$

The first equality in Eq. (3.60) holds trivially, the second can be checked easily using the following properties: $\sigma_x^2 = I$, $\sigma_x R_y(\xi)\sigma_x = R_y(-\xi)$ and $\sigma_x R_z(\xi)\sigma_x = R_z(-\xi)$.

Therefore, it is possible to implement the controlled-U operation as in Fig. 3.8. Indeed, if the value of the control qubit is 0, then $ABC = I$ is applied to the target qubit. If the value of the control qubit is set to 1, then $A\sigma_x B\sigma_x C = \Phi(-\delta)U$ is applied to the target qubit. Therefore, we are close to the implementation of the controlled-U transformation, except for the phase factor $\Phi(-\delta) = e^{-i\delta}I$, which appears when the control is set to 1. The last gate of the circuit in Fig. 3.8 removes this undesired phase factor. It acts non-trivially only on the control qubit and has the following matrix representation:

$$
R_z(\delta) \otimes I =
\begin{bmatrix} 1 & 0 \\ 0 & e^{i\delta} \end{bmatrix} \otimes I =
\begin{bmatrix} 1 \cdot I & 0 \cdot I \\ 0 \cdot I & e^{i\delta} \cdot I \end{bmatrix} =
\begin{bmatrix} 1 & 0 & 0 & 0 \\ 0 & 1 & 0 & 0 \\ 0 & 0 & e^{i\delta} & 0 \\ 0 & 0 & 0 & e^{i\delta} \end{bmatrix},
$$
$$(3.61)$$

where the tensor product of matrices has been performed as explained in Sec. 2.3. This gate is trivially equivalent to the controlled-$\Phi(\delta)$ gate and therefore it removes the undesired phase factor $\Phi(-\delta)$ (as we have seen, this phase factor appears only when the control is set to 1). This completes the proof that the circuit of Fig. 3.8 implements the controlled-U operation.

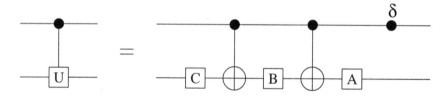

Fig. 3.8   A circuit implementing the controlled-U gate.

We now consider the gate C^k-U, which applies a unitary transformation U to the target qubit if all the k control qubits are set to 1. We shall show that these gates can be implemented by means of elementary gates, namely, single-qubit and CNOT gates.

Of particular interest is C^2-NOT (or *Toffoli gate*), which applies a NOT operation to the target qubit only if the two control qubits are set to 1. The construction of the Toffoli gate is given in Fig. 3.9. Here V is the matrix

$$V = \begin{bmatrix} 1 & 0 \\ 0 & i \end{bmatrix}. \tag{3.62}$$

As we have seen above, since V and $V^\dagger$ are unitary, the operations controlled-V and controlled-$V^\dagger$ can be implemented using only single-qubit gates and CNOT. Thus, these elementary gates are building blocks for the Toffoli gate. This result is of particular importance for the following reasons:

(i) since the Toffoli gate is universal for classical computation (see Chap. 1), quantum circuits having as building blocks single-qubit and CNOT gates encompass classical computation;

(ii) unlike quantum computation, in classical computation one- and two-bit reversible gates are not universal.

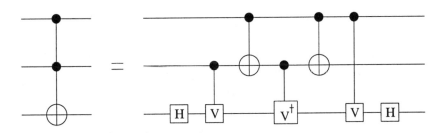

Fig. 3.9   A circuit implementing the Toffoli gate.

Exercise 3.14   Check that for any unitary 2×2 matrix U, the gate C^2-U can be simulated by the circuit in Fig. 3.10, where V is such that $V^2 = U$.

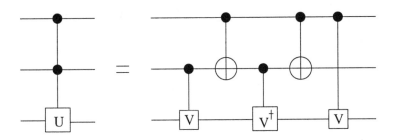

Fig. 3.10   A circuit implementing the C^2-U gate.

The Toffoli gate is also particularly useful in building the C^k-U gate. A simple circuit implementing this gate is shown in Fig. 3.11, for the particular case $k = 4$. It requires $k - 1$ *ancillary* (workspace) qubits, initially set to their $|0\rangle$ state. The first $k - 1$ Toffoli gates change the state of the last ancillary qubit to $|j\rangle$, where j is given by the product $i_{k-1}i_{k-2}i_1i_0$, which is equal to $|1\rangle$ if and only if all the control qubits are initially set to $|1\rangle$. Then a controlled-U operation, having the last ancillary qubit as a control, performs the required C^k-U gate. The last $k - 1$ Toffoli gates refresh the ancillary qubits to their initial state $|0\rangle$.

It is possible to show that C^k-U can be implemented without ancillary qubits (this can be achieved by means of a generalization of the circuit of Fig. 3.10, see Barenco *et al.*, 1995). The price to pay is that the number of elementary gates required is $O(k^2)$ instead of the $O(k)$ elementary gates used in the circuit in Fig. 3.11.

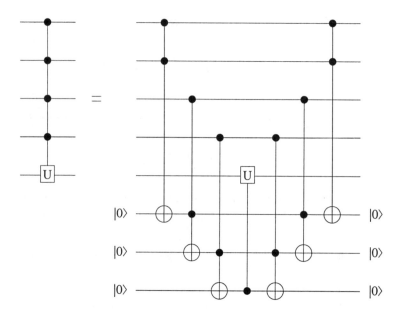

Fig. 3.11 A circuit implementing the C^4-U gate.

The final step in proving that single-qubit and CNOT gates are universal makes use of the decomposition formula (see Barenco *et al.*, 1995)

$$U^{(n)} = \prod_{i=1}^{2^n-1} \prod_{j=0}^{i-1} V_{ij}, \tag{3.63}$$

where $U^{(n)}$ is a generic unitary operator acting on the 2^n-dimensional Hilbert space of n-qubits and V_{ij} induces a rotation of the states $|i\rangle$ and $|j\rangle$ according to a unitary 2×2 matrix. Hence, V_{ij}, when applied to a generic wave vector, acts non-trivially only on two vector components, the component along $|i\rangle$ and the component along $|j\rangle$. The basic idea to implement V_{ij} on a quantum computer is to reduce the rotation of the axis $|i\rangle$ and $|j\rangle$ to a controlled rotation of a single qubit. For this purpose, we write a *Gray code* connecting i and j, namely, a sequence of binary numbers starting with i and finishing with j, whose consecutive members differ in one bit only. For instance, if $i = 00111010$ and $j = 00100111$, we have the Gray code

$$
\begin{aligned}
i &= 0\ 0\ 1\ 1\ 1\ 0\ 1\ 0 \\
  &\ 0\ 0\ 1\ 1\ 1\ 0\ 1\ 1 \\
  &\ 0\ 0\ 1\ 1\ 1\ 1\ 1\ 1 \\
  &\ 0\ 0\ 1\ 1\ 0\ 1\ 1\ 1 \\
j &= 0\ 0\ 1\ 0\ 0\ 1\ 1\ 1
\end{aligned}
\tag{3.64}
$$

Each step of the Gray code can be performed on a quantum computer through a generalized $C^{(n-1)}$-NOT gate. A generalized $C^{(n-1)}$-NOT gate is by definition a gate in which the target qubit is flipped if and only if the $n - 1$ control qubits are in a well-defined state $|i_{n-2} \cdots i_1 i_0\rangle$. Unlike the standard $C^{(n-1)}$-NOT gate, it is not required that this state correspond to $|1 \cdots 11\rangle$. Let us consider, for instance, the first step of the Gray code (3.64). Since it changes $i = 00111010$ into $i' \equiv 00111011$, it can be implemented on a quantum computer by a gate swapping the states $|i\rangle$ and $|i'\rangle$. A generalized C^7-NOT accomplishes this task. It flips the state of the last qubit, provided that the first seven qubits are set to $|0011101\rangle$.

The penultimate line of the Gray code (3.64) differs from j (the last line) in just one bit and therefore the matrix V_{ij} can now be implemented as a rotation of the corresponding qubit, controlled by the states of all the others. Finally, the above permutations are undone in reverse order. At the end of the whole procedure, we have operated only on the states $|i\rangle$ and $|j\rangle$, leaving all the other states unchanged. The quantum circuit implementing the rotation V_{ij} of the states $|i\rangle$ and $|j\rangle$ given in the above example is shown in Fig. 3.12. The action of this circuit can be summarized as follows: to perform a rotation V_{ij} of two generic states $|i\rangle$ and $|j\rangle$, we operate a sequence of state permutations ($|i\rangle = |00111010\rangle \leftrightarrow |i'\rangle = |00111011\rangle$ and so on) up to a final state ($|i_f\rangle = |00110111\rangle$) that differs from $|j\rangle$ only in one qubit. The rotation V_{ij} is then performed on the states $|i_f\rangle$ and $|j\rangle$. Finally, we undo permutations so that $|i_f\rangle$ returns back to the state $|i\rangle$.

This concludes the proof that the single-qubit plus two-qubit CNOT gates are universal gates for quantum computation. Let us recall the main steps of this proof:

(i) for any single-qubit rotation U, the controlled-U operation can be decomposed into single-qubit and CNOT gates;

(ii) the C^2-NOT gate (Toffoli gate) can be implemented using CNOT, controlled-U and Hadamard gates;

(iii) any C^k-U gate ($k > 2$) can be decomposed into Toffoli and controlled-U gates;

(iv) a generic unitary operator $U^{(n)}$ acting on the Hilbert space of n-qubits can be decomposed by means of C^k-U gates.

Exercise 3.15 Show that a generalized $C^{(n-1)}$-NOT gate can be obtained from the standard $C^{(n-1)}$-NOT gate plus single-qubit gates.

The number of elementary gates required to implement the decomposition (3.63) is $O(n^2 4^n)$, since there are $O(2^n \times 2^n = 4^n)$ V-terms in this product and every term requires $O(n^2)$ elementary gates. Actually, each V-term involves at most $2n$ permutations (elements of the Gray code plus its reverse), besides a $C^{(n-1)}$-V_{ij} gate. Each controlled operation requires $O(n)$ elementary gates, provided that one has at ones disposal $n-1$ ancillary qubits (that can be refreshed and reused each time).

We stress that the decomposition described above is in general *not efficient*, i.e., it requires a number of basic operations that scales exponentially with the number of qubits. It is also clear that to implement a generic unitary transformation U involving n qubits, one necessarily needs exponentially many elementary gates, since U is determined by $O(4^n)$ real parameters. A fundamental and still open problem of quantum computation is to discover which special classes of unitary transformations can be computed in the quantum circuit model by means of a *polynomial* number of elementary gates.

It is interesting to give an alternative method (Tucci, 1999) to decom-

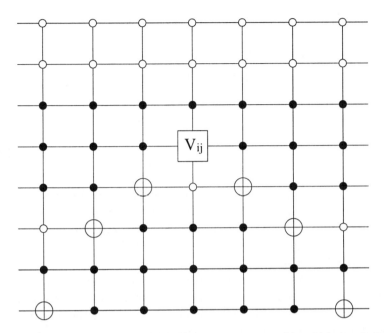

Fig. 3.12   A circuit implementing the rotation V_{ij} of the states $|i\rangle$ and $|j\rangle$ given in (3.64). The circuit uses 6 generalized C^7-NOT gates and a generalized C^7-V_{ij} gate. Empty or full circles indicate that the operation on the target qubit (NOT or V_{ij}) is active only when the control qubits are set to 0 or 1, respectively. Note that the control qubits can be both above and below the target qubit.

pose a generic $2^n \times 2^n$ unitary matrix into a sequence of elementary operations. This method utilizes the CS decomposition (C and S stand for cosine and sine, respectively). Given an $N \times N$ unitary matrix U, where N is an even number, the CS-decomposition theorem tells us that we can always express U in the form

$$U = \begin{bmatrix} L_0 & 0 \\ 0 & L_1 \end{bmatrix} D \begin{bmatrix} R_0 & 0 \\ 0 & R_1 \end{bmatrix}, \tag{3.65}$$

where the matrices L_0, L_1, R_0, and R_1 are $(N/2) \times (N/2)$ unitary matrices and

$$D = \begin{bmatrix} D_C & -D_S \\ D_S & D_C \end{bmatrix}, \tag{3.66}$$

where D_C and D_S are diagonal matrices of the form

$$D_C = \text{diag}(\cos \phi_1, \cos \phi_2, \ldots, \cos \phi_{N/2}),$$
$$D_S = \text{diag}(\sin \phi_1, \sin \phi_2, \ldots, \sin \phi_{N/2}), \tag{3.67}$$

with appropriate angles ϕ_i. It follows from Eq. (3.65) that

$$U = \begin{bmatrix} L_0 D_C R_0 & -L_0 D_S R_1 \\ L_1 D_S R_0 & L_1 D_C R_1 \end{bmatrix}. \tag{3.68}$$

If we take $N = 2^n$, n being the number of qubits, this decomposition can be iterated to matrices of smaller and smaller size. Hence, it is possible to reduce U into a sequence of elementary operations. Note that this decomposition is in general inefficient, since at the end it generates $O(2^n)$ (controlled) 2×2 matrices.

As a simple example, let us consider the decomposition of a 4×4 unitary matrix U. The quantum circuit implementing Eq. (3.65) in this special case is shown in Fig. 3.13, while the decomposition of the matrix D into elementary gates is discussed in exercise 3.16

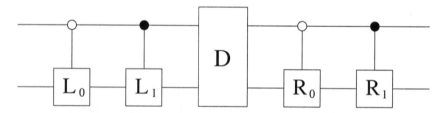

Fig. 3.13   The quantum circuit implementing the decomposition (3.65) of a 4×4 matrix into 4 controlled operations plus a matrix D.

Exercise 3.16   Show that the quantum circuit in Fig. 3.14 implements the unitary matrix (3.66), for a 4×4 matrix D.

Finally, we note that, unlike classical computation, the set of elementary gates that we have introduced is *continuous*. Indeed, this set is composed of CNOT, Hadamard and single-qubit phase-shift gates $R_z(\delta)$. Since δ is a real parameter, phase-shift gates constitute a continuous set. We can easily convince ourselves that the set of elementary gates must be continuous, since the set of unitary transformations in the Hilbert space of n-qubits is

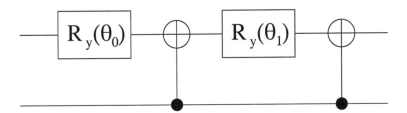

Fig. 3.14 The quantum circuit implementing the matrix D of Eq. (3.66). The rotation matrices $R_y(\theta_i)$ are defined by Eq. (3.33).

continuous. However, it is possible to approximate any such transformation with arbitrary accuracy ϵ using a *discrete* set of quantum gates. For instance, one can use Hadamard, $R_z(\frac{\pi}{4})$ and CNOT gates. Indeed, it is possible to show that the first two gates of this set can approximate any single-qubit rotation with accuracy ϵ in $O(\log^c(1/\epsilon))$ steps, with the constant $c \sim 2$ (see Nielsen and Chuang, 2000). In any case, whatever set of elementary gates, discrete or continuous, is chosen, a generic single-qubit rotation can only be simulated with finite precision by a computer operating with finite resources. For example, we should need an infinite amount of resources in order to exactly specify the real parameter δ in the phase-shift gate $R_z(\delta)$. This naturally raises the question as to the stability of quantum computation in the presence of imperfect unitary operations. We shall address this problem in the next section.

3.5.1 * *Preparation of the initial state*

In this subsection, we discuss the preparation of a generic state of the quantum computer. It turns out that this preparation in general cannot be done efficiently, since it requires a number of gates that is exponential in the number of qubits. Let us assume that the quantum computer is initially in its fiducial state $|0\rangle$ and we wish to prepare the state

$$|\psi\rangle = \sum_{i=0}^{7} a_i |i\rangle, \qquad (3.69)$$

where, for the sake of clarity, we consider the particular case corresponding to $n = 3$ qubits.

Let us first set the amplitudes $|a_i|$ ($a_i = |a_i|e^{i\gamma_i}$). It is easy to check that this is accomplished by the circuit of Fig. 3.15 applied to the state

$|000\rangle$. Indeed, the first gate $(R_y(-2\theta_1))^3$ transforms the state $|000\rangle$ into

$$\big(\cos\theta_1\,|0\rangle + \sin\theta_1\,|1\rangle\big)|00\rangle\,. \tag{3.70}$$

Then the two generalized $C\text{-}R_y$ gates lead to the state

$$\Big(\cos\theta_1\cos\theta_2|00\rangle + \cos\theta_1\sin\theta_2|01\rangle$$
$$+ \sin\theta_1\cos\theta_3|10\rangle + \sin\theta_1\sin\theta_3|11\rangle\Big)|0\rangle\,. \tag{3.71}$$

Finally, the four generalized $C^2\text{-}R_y$ gates generate the state

$$\cos\theta_1\cos\theta_2\cos\theta_4\,|000\rangle + \cos\theta_1\cos\theta_2\sin\theta_4\,|001\rangle$$
$$+ \cos\theta_1\sin\theta_2\cos\theta_5\,|010\rangle + \cos\theta_1\sin\theta_2\sin\theta_5\,|011\rangle$$
$$+ \sin\theta_1\cos\theta_3\cos\theta_6\,|100\rangle + \sin\theta_1\cos\theta_3\sin\theta_6\,|101\rangle$$
$$+ \sin\theta_1\sin\theta_3\cos\theta_7\,|110\rangle + \sin\theta_1\sin\theta_3\sin\theta_7\,|111\rangle\,. \tag{3.72}$$

This state reproduces the amplitudes $|a_i|$ of (3.69), provided that we set the angles θ_i as follows:

$$|a_0| = \cos\theta_1\cos\theta_2\cos\theta_4\,, \qquad |a_1| = \cos\theta_1\cos\theta_2\sin\theta_4\,,$$
$$|a_2| = \cos\theta_1\sin\theta_2\cos\theta_5\,, \qquad |a_3| = \cos\theta_1\sin\theta_2\sin\theta_5\,,$$
$$|a_4| = \sin\theta_1\cos\theta_3\cos\theta_6\,, \qquad |a_5| = \sin\theta_1\cos\theta_3\sin\theta_6\,,$$
$$|a_6| = \sin\theta_1\sin\theta_3\cos\theta_7\,, \qquad |a_7| = \sin\theta_1\sin\theta_3\sin\theta_7\,. \tag{3.73}$$

Thus, the angles θ_i are determined by the amplitudes $|a_i|$ and can be taken in the interval $[0, \frac{\pi}{2}]$.

It is important to note that the circuit of Fig. 3.15 requires $2^n - 1 = 7$ (controlled) single-qubit rotations about the y-axis. This is consistent with the fact that, owing to the normalization constraint, one needs to set $2^n - 1$ degrees of freedom to determine the 2^n wave-function amplitudes.

Now we set the phases γ_i. One has to perform a unitary transformation U_D, whose matrix representation is diagonal in the computational basis $\{|000\rangle, |001\rangle, |010\rangle, |011\rangle, |100\rangle, |101\rangle, |110\rangle, |111\rangle\}$:

$$U_D = \text{diag}[e^{i\gamma_0}, e^{i\gamma_1}, e^{i\gamma_2}, e^{i\gamma_3}, e^{i\gamma_4}, e^{i\gamma_5}, e^{i\gamma_6}, e^{i\gamma_7}]\,. \tag{3.74}$$

---

[3] We remind the reader that the operator R_y was defined in Eq. (3.33) and that $R_y(-2\theta)$ corresponds to a clockwise rotation through an angle 2θ about the y-axis of the Bloch sphere.

It is easy to check that the operator U_D is explicitly constructed in Fig. 3.16 by means of $2^n/2$ controlled operations. In these operations, Γ_k is a single-qubit gate, whose matrix representation in the computational basis $\{|0\rangle, |1\rangle\}$ is given by

$$\Gamma_k = \begin{bmatrix} e^{i\gamma_{2k}} & 0 \\ 0 & e^{i\gamma_{2k+1}} \end{bmatrix}. \tag{3.75}$$

We need $2^N/2$ gates Γ_k, with k ranging from 0 to $2^n/2 - 1$. As can be seen from Fig. 3.16, Γ_0 acts only when the first two qubits are in the state $|00\rangle$ and therefore it sets the phases $e^{i\gamma_0}$ and $e^{i\gamma_1}$ in front of the basis vectors $|000\rangle$ and $|001\rangle$, respectively. Similarly, Γ_1 acts only when the first two qubits are in the state $|01\rangle$ and therefore it sets the phase $e^{i\gamma_2}$ and $e^{i\gamma_3}$ in front of the basis vectors $|010\rangle$ and $|011\rangle$ and so on.

We remark that, while the state preparation in general requires, as in classical computation, an exponential number of operations, the quantum computer has an exponential advantage in memory requirements: A wave vector loaded in n qubits is determined by 2^n complex numbers, the coefficients of its expansion over the computational basis. The classical computer needs $O(2^n)$ bits to load 2^n complex numbers (more precisely, we need $m2^n$ bits, where m is the number of bits required to store a complex number with a given precision). The huge memory capabilities of the quantum computer appear clearly: it accomplishes this task with just n qubits.

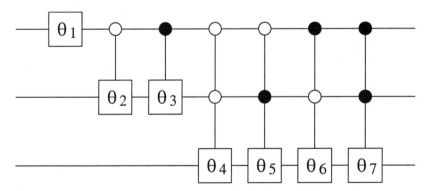

Fig. 3.15 A circuit setting the amplitudes of a generic three-qubit state. The θ_i-symbols stand for the rotation matrices $R_y(-2\theta_i)$ defined by Eq. (3.33).

It is important to realize that, in special cases, a given wave function can be prepared efficiently. We say that an operation in a quantum computer can be performed efficiently if it requires a number of elementary gates polynomial in the number of qubits. For instance, the equal superposition of all states of the computational basis,

$$|\psi\rangle = \sum_{i=0}^{2^n-1} \frac{1}{\sqrt{2^n}} |i\rangle, \tag{3.76}$$

is obtained after the application of n Hadamard gates (one for each qubit) to the state $|0\rangle$.

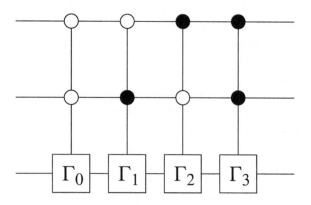

Fig. 3.16   A circuit setting the phases of a generic three-qubit state.

3.6   Unitary errors

Any quantum computation is given by a sequence of quantum gates applied to some initial state:

$$|\psi_n\rangle = \prod_{i=1}^{n} U_i |\psi_0\rangle. \tag{3.77}$$

Since unitary operations form a continuous set, any realistic implementation will involve some error. Let us assume for the time being that errors are unitary, while we delay the discussion of the non-unitary errors due to the unavoidable coupling to the environment until Chap. 6. Instead of the operators U_i, we apply slightly different unitary operators V_i. If we call $|\psi_i\rangle$ the ideal state obtained after i steps, we have

$$|\psi_i\rangle = U_i |\psi_{i-1}\rangle. \tag{3.78}$$

However, if we apply the actual imperfect operation V_i, we obtain

$$V_i |\psi_{i-1}\rangle = |\psi_i\rangle + |E_i\rangle, \tag{3.79}$$

where

$$|E_i\rangle = (V_i - U_i) |\psi_{i-1}\rangle. \tag{3.80}$$

If $|\tilde{\psi}_i\rangle$ denotes the quantum-computer wave function after i imperfect unitary transformations have been applied, then we have

$$\begin{aligned}
|\tilde{\psi}_1\rangle &= |\psi_1\rangle + |E_1\rangle, \\
|\tilde{\psi}_2\rangle &= V_2|\tilde{\psi}_1\rangle = |\psi_2\rangle + |E_2\rangle + V_2|E_1\rangle,
\end{aligned} \tag{3.81}$$

and so on. Therefore, after n iterations we obtain

$$\begin{aligned}
|\tilde{\psi}_n\rangle = {}&|\psi_n\rangle + |E_n\rangle + V_n|E_{n-1}\rangle \\
&+ V_n V_{n-1} |E_{n-2}\rangle + \cdots + V_n V_{n-1} \ldots V_2 |E_1\rangle.
\end{aligned} \tag{3.82}$$

In the worst case the errors are aligned and add linearly. This gives the following bound (a simple consequence of the triangle inequality)

$$\left\| |\tilde{\psi}_n\rangle - |\psi_n\rangle \right\| \leq \left\| |E_n\rangle \right\| + \left\| |E_{n-1}\rangle \right\| + \cdots + \left\| |E_1\rangle \right\|, \tag{3.83}$$

where we have used the fact that the evolution has been assumed unitary:

$$\left\| V_i |E_{i-1}\rangle \right\| = \left\| |E_{i-1}\rangle \right\|. \tag{3.84}$$

We can bound the Euclidean norm of the error vector $|E_i\rangle$ as follows:

$$\left\| |E_i\rangle \right\| = \left\| (V_i - U_i) |\psi_{i-1}\rangle \right\| \leq \left\| V_i - U_i \right\|_{\text{sup}}, \tag{3.85}$$

where $\left\| V_i - U_i \right\|_{\text{sup}}$ is the sup norm of the operator $V_i - U_i$, that is, its eigenvalue of maximum modulus. Assuming that the error is uniformly bound at each step,

$$\left\| V_i - U_i \right\|_{\text{sup}} < \epsilon, \tag{3.86}$$

we obtain after the application of n imperfect operators

$$\left\| |\tilde{\psi}_n\rangle - |\psi_n\rangle \right\| < n\epsilon. \tag{3.87}$$

Therefore, unitary errors accumulate at worst linearly with the length of the quantum computation. This growth takes place for systematic errors that line up in the same direction, while stochastic errors are randomly directed and therefore give a more favourable $\sqrt{n}$ growth.

We note that, if a quantum computation requires n elementary gates (single-qubit and CNOT gates), it can be approximated to accuracy ϵ using $O(n \log^c(1/(\epsilon/n)))$ gates from the discrete set introduced in the previous section (Hadamard, $R_z(\frac{\pi}{4})$ and CNOT). Indeed, as we stated in that section, any single-qubit rotation can be approximated with accuracy ϵ/n in $O(\log^c(1/(\epsilon/n)))$ gates from the above discrete set and the bound of Eq. (3.87) implies that it is sufficient to improve the accuracy of the gates linearly with the length n of the quantum computation.

It is important to connect the accuracy of the quantum-computer wave function (measured by $\| |\tilde{\psi}_n\rangle - |\psi_n\rangle \|$) with the accuracy of the results of a quantum computation. Quantum computation ends up with a projective measurement in the computational basis, giving outcome i with probability $p_i = |\langle i|\psi_n\rangle|^2$. In the presence of unitary errors, the real probability becomes $\tilde{p}_i = |\langle i|\tilde{\psi}_n\rangle|^2$. It is possible to show that $\sum_i |p_i - \tilde{p}_i| \leq 2\| |\tilde{\psi}_n\rangle - |\psi_n\rangle \|$ (see Preskill, 1998).

Finally, we note that the argument developed in this section does not allow us to determine how errors scale with the number of qubits in the quantum computer.

3.7 Function evaluation

The basic task performed by a classical computer is the evaluation of a binary (logic) function with an n-bit input and a one-bit output:

$$f : \{0,1\}^n \rightarrow \{0,1\}. \tag{3.88}$$

This means that f takes the input $(x_{n-1}, \ldots, x_1, x_0)$, with $x_{n-1}, \ldots, x_1, x_0$ binary digits, and produces the output $f(x_{n-1}, \ldots, x_1, x_0)$, which can be equal to 0 or 1. A computer can evaluate any complicated function by assembling these binary functions (see Chap. 1). In this section, we shall discuss the implementation of such functions in a quantum computer.

Let us first discuss the construction of the binary functions for $n = 2$ bits. There are $2^{2^n} = 16$ two-bit logic functions, which we show in Table 3.1. These functions are not in general invertible. For instance, $f_1(x_1, x_0) = x_1 \wedge x_0$ ($\wedge$ denotes the logic AND) is equal to 0 for three different inputs. As we saw in Chap. 1, these functions can be evaluated in reversible computation if an ancillary bit is added. The appropriate unitary transformation that evaluates the function f on a quantum computer makes use of the ancillary qubit $|y\rangle$ and is given by

$$U_f|x_{n-1}, x_{n-2}, \ldots, x_0\rangle|y\rangle$$
$$= |x_{n-1}, x_{n-2}, \ldots, x_0\rangle|y \oplus f(x_{n-1}, x_{n-2}, \ldots, x_0)\rangle, \tag{3.89}$$

in which, for a given output, there is a unique input.

Exercise 3.17 Show that the transformation (3.89) is unitary.

The explicit construction of the binary functions of Table 3.1 is given by the quantum circuits of Fig. 3.17. We show only 8 functions, since $f_{15-i} = \overline{f}_i$, where the bar indicates NOT. Therefore, f_{15-i} can be obtained from f_i simply by application of a NOT (σ_x) gate to the ancillary qubit. We note that the function $f_6 = x_1 \oplus x_0$ could be implemented reversibly by means of a CNOT gate, with no need of any ancillary qubit. The same holds for $f_3 = x_1$ and $f_5 = x_0$, which simply correspond to the input value of one of the bits, while $f_0 = 0$ is a fully degenerate constant function.

Table 3.1   Two-bit logic functions.

$x_1 x_0$	f_0	f_1	f_2	f_3	f_4	f_5	f_6	f_7	f_8	f_9	f_{10}	f_{11}	f_{12}	f_{13}	f_{14}	f_{15}
0 0	0	0	0	0	0	0	0	0	1	1	1	1	1	1	1	1
0 1	0	0	0	0	1	1	1	1	0	0	0	0	1	1	1	1
1 0	0	0	1	1	0	0	1	1	0	0	1	1	0	0	1	1
1 1	0	1	0	1	0	1	0	1	0	1	0	1	0	1	0	1

Now let us consider a binary function with a generic number n of input bits. As we saw in Chap. 1, one way of expressing a binary function $f(x)$ $(x = (x_{n-1}, x_{n-2}, \ldots, x_1, x_0))$ is to consider its *minterms*, defined, for each $x^{(a)}$ such that $f(x^{(a)}) = 1$, as

$$f^{(a)}(x) = \begin{cases} 1 & \text{if } x = x^{(a)}, \\ 0 & \text{otherwise.} \end{cases} \tag{3.90}$$

Then the function $f(x)$ is written

$$f(x) = f^{(1)}(x) \vee f^{(2)}(x) \vee \cdots \vee f^{(m)}(x), \tag{3.91}$$

where f is the logic $\vee$ (OR) of all $0 \leq m \leq 2^n$ minterms. Note that in Eq. (3.91) we need one minterm for each x value such that $f(x) = 1$. It is sufficient to compute the minterms in order to obtain $f(x)$.

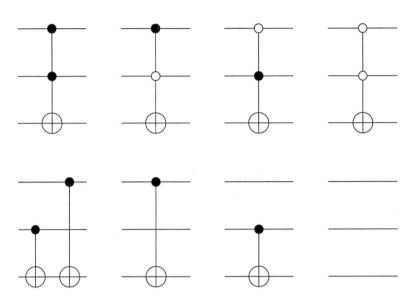

Fig. 3.17 Quantum circuits implementing the two-bit binary functions. In each circuit the three lines represent (from top to bottom) the more significant qubit, the less significant one and the ancillary, which is input in the state $|0\rangle$. From left to right: $f_1 = x_1 \wedge x_0$, $f_2 = x_1 \wedge \bar{x}_0$, $f_4 = \bar{x}_1 \wedge x_0$, $f_8 = \bar{x}_1 \wedge \bar{x}_0$ (top); $f_6 = x_1 \oplus x_0$, $f_3 = x_1$, $f_5 = x_0$, and $f_0 = 0$ (bottom).

Each minterm is implemented in a quantum computer by a generalized C^n-NOT gate. We remark that for a generic function with no structure, the number m of minterms grows exponentially with n and there is no way to evaluate f efficiently (*i.e.*, with a number of elementary gates polynomial in n).

We now give an example of function evaluation, for the binary function $f(x_2, x_1, x_0)$ defined by the truth table of Table 3.2. There are three minterms, for $x^{(1)} = (0,0,1)$, $x^{(2)} = (1,0,0)$ and $x^{(3)} = (1,0,1)$. The corresponding quantum circuit implementing the evaluation of the function f is shown in Fig. 3.18. In this circuit, each generalized C^3-NOT gate corresponds to a minterm. Since the minterms $x^{(2)}$ and $x^{(3)}$ differ only in the value x_0 of the third bit, it is possible to simplify the circuit of Fig. 3.18, with a generalized C^2-NOT gate (controlled by x_2, x_1) instead of two generalized C^3-NOT gates (controlled by x_2, x_1, x_0). This reflects the logic identity $x_0 + \bar{x}_0 = 1$. We point out that the design of optimized circuits is a basic problem of computation: simplification rules of quantum logic circuits are given in Lee *et al.* (1999).

Table 3.2 An example of
a truth table for a binary
function.

x_2	x_1	x_0	f
0	0	0	0
0	0	1	1
0	1	0	0
0	1	1	0
1	0	0	1
1	0	1	1
1	1	0	0
1	1	1	0

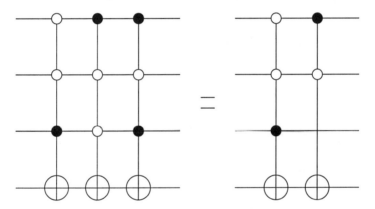

Fig. 3.18 A quantum circuit implementing the binary function f defined in Table 3.2: decomposition of f in minterms (left) and the simplified circuit (right). The input for the four qubits is (from top to bottom) $|x_2\rangle$, $|x_1\rangle$, $|x_0\rangle$ and $|0\rangle$.

Let us consider a further example, the computation of the function x^2 for a 2-qubit input. In general, if we need $n = \log_2 N$ qubits to load an integer $x \in [1, N]$, we need $2n = \log_2 N^2$ qubits to load $x^2 \in [1, N^2]$. Thus, the case $n = 2$ requires $2n = 4$ qubits to load the output. This corresponds to 4 binary functions, which can be evaluated reversibly using 4 ancillary ancillary qubits. The truth table for the binary function x^2 is shown in Table 3.3, while the corresponding quantum circuit is drawn in Fig. 3.19. We note that in this circuit one of the ancillary qubits is never addressed, since it gives the constant binary function $f_0 = 0$.

Finally, we emphasize that what makes quantum function evaluation

Table 3.3 The truth table for the function $f(x) = x^2$ (2-bit input).

x_1	x_0	x	x^2	x^2
0	0	0	0	0000
0	1	1	1	0001
1	0	2	4	0100
1	1	3	9	1001

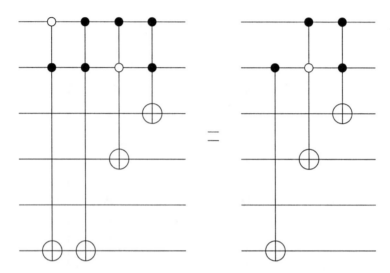

Fig. 3.19 A quantum circuit implementing the function $f(x) = x^2$ for $n = 2$-bit input: decomposition in minterms (left) and the simplified circuit (right). The two top lines represent the input x, the four bottom lines the ancillary qubits, which are input in the state $|0000\rangle$. Their output loads x^2.

interesting is its action on an input state given by the superposition of *exponentially many* states of the computational basis. Owing to the linearity of quantum mechanics we have:

$$U_f \sum_{x=1}^{2^n-1} c_x |x\rangle |y\rangle = \sum_{x=1}^{2^n-1} c_x |x\rangle |y \oplus f(x)\rangle, \qquad (3.92)$$

which produces $f(x)$ for all x in a single run. For instance, if we consider the function $f(x) = x^2$ and the input

$$\tfrac{1}{2}\big(|0\rangle + |1\rangle + |2\rangle + |3\rangle\big)|0\rangle, \qquad (3.93)$$

a single run of the circuit in Fig. 3.19 produces the output

$$\tfrac{1}{2}\big(|0\rangle|0\rangle + |1\rangle|1\rangle + |2\rangle|4\rangle + |3\rangle|9\rangle\big)\,. \tag{3.94}$$

Thus, we have computed in parallel $f(x) = x^2$ for $x = 0$, 1, 2 and 3, a possibility that is beyond the reach of the classical computer, which can only receive as input a given value of x, not a superposition of x values. This distinctive property of the quantum computer is known as *quantum parallelism*.

However, it is not an easy task to extract useful information from the superposition (3.92). The problem is that this information is, in a sense, hidden. A projective measurement of the first register in the computational basis (that is, we measure the qubit polarization along the z-axis for all the qubits in this register) yields a particular value x', after which a measurement of the second register will necessarily give outcome $f(x')$. For example, if we measure the first register of the wave function (3.94), we obtain with equal probabilities $p_0 = p_1 = p_2 = p_3 = \tfrac{1}{4}$ the four possible outcomes 0, 1, 2 and 3. Let us assume that the outcome is $x' = 2$. Then the state (3.94) collapses onto the state $|2\rangle|4\rangle$. Therefore, a measurement of the second register now gives outcome $f(x') = 4$ with unit probability. Hence, we end up with the evaluation of the function $f(x)$ for a single value of x, exactly as with a classical computer. However, over the next sections we shall discuss quantum algorithms that exploit quantum interference to *extract* efficiently from the superposition state (3.92) information other than just a single value of f.

3.8 The quantum adder

As in classical computation, it is important to construct quantum circuits for performing basic elementary operations. Explicit constructions for several operations, including plain addition, modular addition and modular exponentiation, can be found in Vedral *et al.* (1996). Here we describe only the plain addition of two n-bit integers a and b (in binary representation, $a = a_{n-1}2^{n-1} + a_{n-2}2^{n-2} + \cdots + a_0 \equiv a_{n-1}a_{n-2}\cdots a_0$ and analogously for $b \equiv b_{n-1}b_{n-2}\cdots b_0$). Following the previous section, one could consider $n + 1$ ancillary qubits and compute reversibly

$$|a, b, 0\rangle \to |a, b, a + b\rangle\,. \tag{3.95}$$

Since we have a $2n$-qubit input (encoding a and b) and the output $a + b$ needs $n + 1$ bits to be encoded without overflows, we must evaluate $n + 1$ binary functions with $2n$-bit input. This procedure is not convenient. It is more useful to compute bit by bit the following:

$$|a, b\rangle \rightarrow |a, a + b\rangle. \tag{3.96}$$

Since the input can be reconstructed from the output, this computation can be performed reversibly.

The sum is performed starting from the least significant qubit, which is the usual way to perform additions in classical computation. For the qubit i, given a_i, b_i and the carry c_i, we need to compute the sum $s_i = a_i \oplus b_i \oplus c_i$ and the new carry c_{i+1}. Therefore, the evaluation of two 3-bit input binary functions is required, one is called SUM and computes s_i, the other is called CARRY and computes c_{i+1}. The corresponding truth tables are given in Table 3.4. The function $\text{SUM}(a_i, b_i, c_i)$ gives

$$s_i = a_i \oplus b_i \oplus c_i, \tag{3.97}$$

while $\text{CARRY}(a_i, b_i, c_i)$, as can be readily checked, produces

$$c_{i+1} = (a_i \wedge b_i) \vee (c_i \wedge a_i) \vee (c_i \wedge b_i). \tag{3.98}$$

Thus, $c_{i+1} = 1$ when at least two of the input bits a_i, b_i and c_i are set to one. The function SUM can be computed directly from the expression (3.97) by means of two CNOT gates. In contrast, the function CARRY involves irreversible logic functions (AND, OR) and therefore requires an ancillary qubit. The quantum circuits implementing SUM and CARRY are shown in Fig. 3.20.

Table 3.4 Truth tables for the functions SUM and CARRY.

c_i	a_i	b_i	s_i	c_{i+1}
0	0	0	0	0
0	0	1	1	0
0	1	0	1	0
0	1	1	0	1
1	0	0	1	0
1	0	1	0	1
1	1	0	0	1
1	1	1	1	1

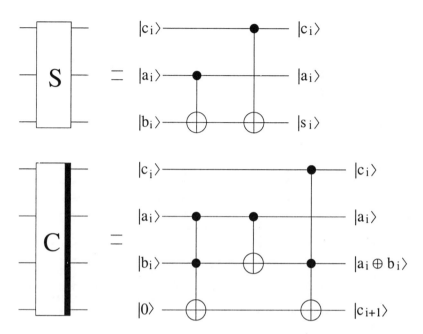

Fig. 3.20   The quantum circuit implementing the functions SUM (upper) and CARRY (lower).

The circuits implementing SUM and CARRY are the building blocks of the plain adder circuit, shown in Fig. 3.21, which computes the sum $a + b$. We need three registers: the first (n qubits) with input and output $|a_{n-1}, a_{n-2}, \ldots, a_1, a_0\rangle$, the second ($n + 1$ qubits) with input $|0, b_{n-1}, b_{n-2}, \ldots, b_1, b_0\rangle$ and output $|(a+b)_n, (a+b)_{n-1}, (a+b)_{n-2}, \ldots, (a+b)_1, (a+b)_0\rangle$, the third consists of $n - 1$ qubits, initially in the state $|0\rangle$, to which the carries are temporarily written and which are refreshed to the $|0\rangle$ state at the end. Each of the first n CARRY's of the circuit transforms $|c_i, a_i, b_i, 0\rangle$ into $|c_i, a_i, a_i \oplus b_i, c_{i+1}\rangle$. The last carry gives the most significant digit of the sum, $(a + b)_n$. Then a single CNOT gate takes $(a_{n-1}, a_{n-1} \oplus b_{n-1})$ into (a_{n-1}, b_{n-1}) and a SUM operation takes $(c_{n-1}, a_{n-1}, b_{n-1})$ into $(c_{n-1}, a_{n-1}, a_{n-1} + b_{n-1})$. After this we apply CARRY and SUM $n - 1$ times. As a result, for each qubit i the sum $a_i + b_i$ is computed while every ancillary qubit is restored to its initial state $|0\rangle$. This is important since it allows us to use these ancillary qubits again for other computations.

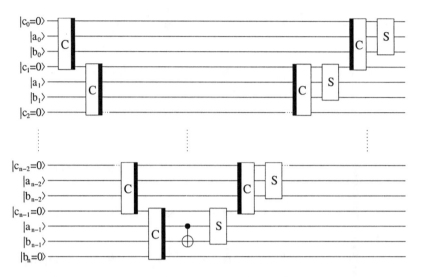

Fig. 3.21   A quantum circuit implementing the addition of two n-bit integers a and b. Note the position of the thick black bars on the CARRY circuits. A CARRY with the bar on the left represents the reverse sequence of elementary gates with respect to a CARRY with the bar on the right.

3.9   Deutsch's algorithm

Deutsch's problem illustrates the computational power of quantum *interference*. We consider a *black box* (called the *oracle*) evaluating a one-bit Boolean function $f : \{0,1\} \rightarrow \{0,1\}$. There are 4 such functions, shown in Table 3.5. They differ in the following *global* property: two of them are *constant* (f_0 and f_3) and two *balanced* (f_1 and f_2). The problem is to decide whether a given function is constant or balanced. The solution to this problem necessarily requires two queries of the oracle in a classical computer. We shall show in this section that a quantum computer can solve the same problem with only one oracle query.

Table 3.5   One-bit logic functions.

x	f_0	f_1	f_2	f_3
0	0	0	1	1
1	0	1	0	1

The quantum circuit implementing Deutsch's algorithm is shown in Fig. 3.22. The function $f(x)$ is evaluated in reversible computation using an ancillary qubit $|y\rangle$. The unitary transformation U_f transforms $|x\rangle|y\rangle$ into $|x\rangle|y \oplus f(x)\rangle$; that is, it flips the second qubit if and only if $f(x) = 1$. The initial state of the qubits is $|x\rangle|y\rangle = |0\rangle|1\rangle$. Then a Hadamard gate prepares the first qubit in the superposition $(|0\rangle + |1\rangle)/\sqrt{2}$. As will be shown below, this will allow the quantum computer to evaluate both $f(0)$ and $f(1)$ in a single run, a possibility that is beyond the reach of a classical computer. Another Hadamard gate prepares the ancillary qubit in $(|0\rangle - |1\rangle)/\sqrt{2}$. This is crucial since, for each $x \in \{0, 1\}$,

$$U_f|x\rangle \tfrac{1}{\sqrt{2}}(|0\rangle - |1\rangle) = (-1)^{f(x)}|x\rangle \tfrac{1}{\sqrt{2}}(|0\rangle - |1\rangle), \qquad (3.99)$$

namely, the phase factor $(-1)^{f(x)}$ is propagated backwards (*kicked back*) in front of the first qubit. Thus, the state of the quantum computer after the function evaluation is

$$\tfrac{1}{\sqrt{2}}\left[(-1)^{f(0)}|0\rangle + (-1)^{f(1)}|1\rangle\right] \tfrac{1}{\sqrt{2}}(|0\rangle - |1\rangle). \qquad (3.100)$$

The second qubit is no longer used and from now on we shall ignore it. The final Hadamard gate leaves the first qubit in the state

$$\tfrac{1}{2}\left\{\left[(-)^{f(0)} + (-)^{f(1)}\right]|0\rangle + \left[(-)^{f(0)} - (-)^{f(1)}\right]|1\rangle\right\}. \qquad (3.101)$$

If $f(0) = f(1)$, this state is equal to $|0\rangle = |f(0) \oplus f(1)\rangle$. If instead $f(0) \neq f(1)$, this state is $|1\rangle = |f(0) \oplus f(1)\rangle$ (in both cases up to an overall phase factor of no physical significance). In any case, we can write the final state of the first qubit as

$$|f(0) \oplus f(1)\rangle. \qquad (3.102)$$

Then a measurement of the first qubit gives with unit probability the outcome 0 if the function is constant and the outcome 1 if the function is balanced. Therefore, a global property of the function $f(x)$ has been encoded in a single qubit after a single call of f. This is because a quantum computer can evaluate both $f(0)$ and $f(1)$ simultaneously. The main point is that these two alternatives "paths" are combined by the final Hadamard gate, giving the desired interference pattern. The interference is constructive for the outcome $f(0) \oplus f(1)$ and destructive for the alternative outcome.

3.9.1 *The Deutsch–Jozsa problem*

Now we shall consider some generalizations of Deutsch's problem. The Deutsch–Jozsa algorithm solves the following problem in a single oracle

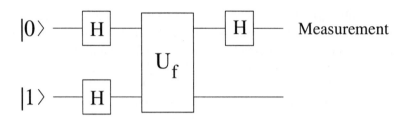

Fig. 3.22   A quantum circuit implementing Deutsch's algorithm.

query: we have an n-bit binary function $f : \{0,1\}^n \to \{0,1\}$, which is "promised" to be constant or balanced (*i.e.*, it has an equal number of output 0's and 1's), and we wish to determine if it is constant or balanced. The quantum circuit that solves this problem is the same as for Deutsch's algorithm (Fig. 3.22), but with n qubits to store the input $x = (x_{n-1}, x_{n-2}, \ldots, x_0)$. The Hadamard gates are now applied in parallel to all n qubits,

$$H^{\otimes n} = H \otimes H \otimes \cdots \otimes H. \qquad (3.103)$$

It is easy to check that the action of $H^{\otimes n}$ on a state $|x\rangle$ of the computational basis gives

$$H^{\otimes n}|x\rangle = \prod_{i=0}^{n-1} \left(\frac{1}{\sqrt{2}} \sum_{y_i=0}^{1} (-1)^{x_i y_i} |y_i\rangle \right) = \frac{1}{2^{n/2}} \sum_{y=0}^{2^n-1} (-1)^{x \cdot y} |y\rangle, \quad (3.104)$$

where $x \cdot y$ denotes the inner product of x and y, modulo 2:

$$x \cdot y = x_{n-1} y_{n-1} \oplus x_{n-2} y_{n-2} \oplus \cdots \oplus x_0 y_0. \qquad (3.105)$$

The circuit in Fig. 3.22, generalized to n-bit input, applies the transformation

$$(H^{\otimes n} \otimes I) U_f (H^{\otimes n} \otimes H) \qquad (3.106)$$

to the input $|00 \ldots 0\rangle |1\rangle$ and generates the output

$$\left(\frac{1}{2^n} \sum_{x,y=0}^{2^n-1} (-1)^{f(x)+x \cdot y} |y\rangle \right) \frac{1}{\sqrt{2}} \left(|0\rangle - |1\rangle \right). \qquad (3.107)$$

This implies that, at the end of the circuit, a measurement of the n qubits in the computational basis gives the state $|00 \ldots 0\rangle$ with unit probability if f is constant and with zero probability if f is balanced. Therefore, a single run of the algorithm, with a single query of the function $f(x)$, determines with certainty if f is constant or balanced. This is an impressive result, since in classical computation one can be sure that $f(x)$ is balanced only after $2^n/2+1$ queries of the function. However, it would be rather unfair to claim an exponential gain of the quantum computer in this algorithm. Indeed, if f is balanced, the probability of obtaining the same response every time is $1/2^{k-1}$, with k number of function queries. Since this probability drops exponentially with k, in this case one can guess that f is constant with a probability of giving the wrong answer that drops exponentially with k. Therefore, as we have discussed in Sec. 1.3, a classical algorithm can reduce the probability of error below a level ϵ after a number of queries $k = O(\log(1/\epsilon))$. We also note that on physical grounds it is not reasonable to ask the quantum computer to give an answer with absolute certainty, since some amount of error will be unavoidably present (see Sec. 3.6 and Chap. 6).

3.9.2   * An extension of Deutsch's algorithm

As an exercise, we now consider a further generalization of the algorithm, which shows that other global properties of a given function can be determined using a quantum computer. Let us consider, for the sake of simplicity, the case of $n = 2$ qubits. The state of the 2 qubits at the end of the Deutsch–Jozsa circuit can be written in the computational basis as $|\psi_f\rangle = c_0|00\rangle + c_1|01\rangle + c_2|10\rangle + c_3|11\rangle$, with the amplitudes c_i shown in Table 3.6 for the 16 two-qubit logic functions of Table 3.1 (we omit the normalization factor to simplify writing).

Table 3.6   Two-qubit states at the end of the Deutsch–Jozsa circuit.

	f_0	f_1	f_2	f_3	f_4	f_5	f_6	f_7	f_8	f_9	f_{10}	f_{11}	f_{12}	f_{13}	f_{14}	f_{15}
c_0	4	2	2	0	2	0	0	-2	2	0	0	-2	0	-2	-2	-4
c_1	0	2	-2	0	2	4	0	2	-2	0	-4	-2	0	2	-2	0
c_2	0	2	2	4	-2	0	0	2	-2	0	0	2	-4	-2	-2	0
c_3	0	-2	2	0	2	0	4	2	-2	-4	0	-2	0	-2	2	0

The measurement of the two qubits gives the outcome 00 with unit probability for the constant functions f_0 and f_{15}, and with zero probability for the balanced functions f_3, f_5, f_6, f_9, f_{10} and f_{12}. Furthermore, if the function is balanced, the measurement allows one to distinguish between three subclasses, that is, f_5 and f_{10} (if the outcome is 01), f_3 and f_{12} (outcome 10) and f_6 and f_9 (outcome 11). We note that it is not possible to distinguish between f_i and f_{15-i}, since the global sign of the wave function is not observable. With the Deutsch–Jozsa circuit, it is not possible to learn anything about the remaining 8 two-bit logic functions (f_1, f_2, f_4, f_7, f_8, f_{11}, f_{13} and f_{14}) since the 4 possible outcomes of the two-qubit measurement have equal probability.

It is also possible to discriminate between the other global properties of the two-qubit binary functions by performing a permutation of the two-qubit binary functions f_i. These permutations can be implemented conveniently if the transformation (3.106) is slightly modified and becomes:

$$\left(H^{\otimes 2} A_j \otimes I\right) U_f \left(H^{\otimes 2} \otimes H\right), \tag{3.108}$$

where the A_j are unitary diagonal matrices, with diagonal elements equal to ± 1. There are $2^{2^n} = 16$ such matrices, which can be coded similarly to the Boolean function f_i:

$$A_0 \equiv \mathrm{diag}\{1,1,1,1\}, \qquad A_1 \equiv \mathrm{diag}\{1,1,1,-1\}, \quad \ldots \tag{3.109}$$

It is easy to check that each unitary transformation A_j changes $f(x)$ in the output wave function (3.107) as follows: $f(x) \to f(x) \oplus (A_j)_{xx}$, where $(A_j)_{xx} = \langle x|A_j|x \rangle$ denotes a diagonal matrix element of A_j. Thus, each unitary transformation A_j induces a permutation of the functions f_i. Therefore, the standard Deutsch–Jozsa circuit (*i.e.*, $A_j = I = A_0$) allows us to discriminate with certainty between

$$\left\{(f_0, f_{15}), (f_3, f_{12}), (f_5, f_{10}), (f_6, f_9)\right\}, \tag{3.110}$$

while by using A_1 we can discriminate between

$$\left\{(f_1, f_{14}), (f_2, f_{13}), (f_4, f_{11}), (f_7, f_8)\right\}, \tag{3.111}$$

and so on.

3.10 Quantum search

In this section, we show that a quantum computer can usefully face the following problem: searching for one marked item inside an *unstructured* database of $N = 2^n$ items. Let us give a simple example: we have a phone book and a given number and we wish to find the corresponding name. The best we can do with a classical computer is to go through the phone book, until we find the solution (the name). This means that it is easy to recognize the solution, but difficult to find it. This is a characteristic of the problems in the computational class **NP** and in many cases to solve these problems there is no better classical algorithm than exhaustive search through all possible solutions. In the following, we shall show that a quantum computer considerably speeds up this search.

The search problem can be rephrased as an *oracle problem*: we label the items of the database as $\{0, 1, \ldots, N-1\}$ and x_0 is the unknown marked item. The oracle computes the n-bit binary function

$$f : \{0,1\}^n \rightarrow \{0,1\}, \tag{3.112}$$

defined as

$$f(x) = \begin{cases} 1 & \text{if } x = x_0, \\ 0 & \text{otherwise.} \end{cases} \tag{3.113}$$

The problem is to find x_0 with the minimum number of queries of the oracle f. Elementary probability theory tells us that, if one enters k items into the black-box function f, the probability of finding x_0 is k/N. Therefore, in order to find x_0 with success probability p, each classical algorithm requires $pN = O(N)$ oracle queries. Grover showed that the same problem can be solved by a quantum computer in $O(\sqrt{N})$ queries. Thus, the quantum computer allows a quadratic speed up. The gain is not exponential and therefore does not change the computational class of the problem, but still there is a significant improvement. To give an idea of the importance of a quadratic speed up, it is sufficient to consider an example from classical computation, the fast Fourier transform algorithm. Indeed, the gain of the fast Fourier transform over the standard Fourier transform is quadratic and the fast Fourier transform has had an enormous impact on signal analysis and countless other applications.

3.10.1 *Searching one item out of four*

It is instructive to examine first the simple case of finding one item out of $N = 4$ items (two qubits). The quantum circuit corresponding to this simple instance of Grover's algorithm is shown in Fig. 3.23. Initially, the two qubits are prepared in the state $|00\rangle$, while the required ancillary qubit is in the state $|y\rangle = |1\rangle$. The first Hadamard gates prepare the two qubits in the equal superposition state and the ancillary in the state $\frac{1}{\sqrt{2}}(|0\rangle - |1\rangle)$.

Thus, the quantum-computer wave function becomes

$$\tfrac{1}{2}\big(|00\rangle + |01\rangle + |10\rangle + |11\rangle\big)\tfrac{1}{\sqrt{2}}\big(|0\rangle - |1\rangle\big)\,. \qquad (3.114)$$

Then we query the oracle and evaluate the function $f(x)$. Indeed, after the oracle query we have $|x\rangle|y\rangle \rightarrow |x\rangle|y \oplus f(x)\rangle$. This means that the value $f(x)$ provided by the oracle is loaded into the ancillary qubit. Since the ancillary has been prepared in the state $\tfrac{1}{\sqrt{2}}(|0\rangle - |1\rangle)$, it stays the same when $f(x) = 0$ and changes its sign when $f(x) = 1$. In the following we consider the special case in which the oracle produces $f(x) = 1$ only when $x = x_0 = (1,0)$. There is no loss of generality in this since the other possible x_0 values are treated equivalently. The circuit drawn in Fig. 3.23 solves the searching problem for the four possible values of x_0. Therefore, the quantum-computer wave function after the oracle query is given by

$$\tfrac{1}{2}\big(|00\rangle + |01\rangle - |10\rangle + |11\rangle\big)\tfrac{1}{\sqrt{2}}\big(|0\rangle - |1\rangle\big)\,, \qquad (3.115)$$

which differs from the state (3.114) in the sign of the coefficient in front of the marked state. Note that, similarly to Deutsch's algorithm, the sign has been kicked back in front of the register $|x\rangle$. The second register is unchanged and we do not consider it any more.

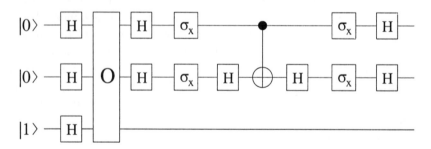

Fig. 3.23   A quantum circuit implementing Grover's algorithm for $N = 4$ items. The Pauli matrix σ_x performs the NOT gate. The rectangle with a letter O inside denotes the oracle query.

After the oracle query, a measurement of the $|x\rangle$ register would not distinguish between the different states of the computational basis, since the amplitudes of the coefficients in front of them are all the same. The key point of Grover's algorithm is to transform the phase difference, which appears in front of the component $|10\rangle$ in (3.115), into an amplitude difference.

This is achieved by means of the following unitary transformation:

$$D_{ij} = -\delta_{ij} + \frac{2}{2^n}. \tag{3.116}$$

In the present case with $n = 2$, the matrix D is given by

$$D = \frac{1}{2} \begin{bmatrix} -1 & 1 & 1 & 1 \\ 1 & -1 & 1 & 1 \\ 1 & 1 & -1 & 1 \\ 1 & 1 & 1 & -1 \end{bmatrix}. \tag{3.117}$$

The transformation D can be implemented by means of that part of the circuit in Fig. 3.23 that follows the oracle query. Indeed, this transformation can be decomposed as follows:

$$D = H^{\otimes 2} D' H^{\otimes 2}, \tag{3.118}$$

where the diagonal matrix

$$D' = \begin{bmatrix} 1 & 0 & 0 & 0 \\ 0 & -1 & 0 & 0 \\ 0 & 0 & -1 & 0 \\ 0 & 0 & 0 & -1 \end{bmatrix} \tag{3.119}$$

gives a controlled phase shift through an angle π (a minus sign) in the coefficient in front of the basis element $|00\rangle$. The matrix D' can be decomposed (up to an overall phase) in the following way:

$$D' = \sigma_x^{\otimes 2} (I \otimes H) \, \mathrm{CNOT} \, (I \otimes H) \sigma_x^{\otimes 2}, \tag{3.120}$$

where $\sigma_x^{\otimes 2} = \sigma_x \otimes \sigma_x$. Indeed, $(I \otimes H) \, \mathrm{CNOT} \, (I \otimes H) = \mathrm{CMINUS}$ (see exercise 3.10). In addition, the NOT gates $\sigma_x^{\otimes 2}$ allow the phase factor to be placed in front of the state $|00\rangle$ instead of the state $|11\rangle$, as it would in a standard controlled phase-shift gate.

The application of the transformation D to the state of Eq. (3.115) gives

$$\mathbf{D} \frac{1}{2} \begin{bmatrix} 1 \\ 1 \\ -1 \\ 1 \end{bmatrix} = \frac{1}{4} \begin{bmatrix} -1 & 1 & 1 & 1 \\ 1 & -1 & 1 & 1 \\ 1 & 1 & -1 & 1 \\ 1 & 1 & 1 & -1 \end{bmatrix} \begin{bmatrix} 1 \\ 1 \\ -1 \\ 1 \end{bmatrix} = \begin{bmatrix} 0 \\ 0 \\ 1 \\ 0 \end{bmatrix}. \tag{3.121}$$

Thus, a standard measurement of the two qubits now gives outcome 10 with certainty. Therefore, the problem has been solved with a single query of the function f, while a classical computer requires, on average, $N_c = 2.25$

queries (the marked item can be found after the first, second or third query, with probability $\frac{1}{4}$ each time; after the third function evaluation the fourth is in any case useless since we know that there is only one marked item and therefore $N_c = \frac{1}{4} \times 1 + \frac{1}{4} \times 2 + \frac{1}{2} \times 3 = 2.25$).

3.10.2 *Searching one item out of N*

Now we describe Grover's algorithm for a generic number of items $N = 2^n$. Again we need a single ancillary qubit $|y\rangle$ and we prepare the quantum-computer wave function in the state $|x\rangle|y\rangle = |00\ldots0\rangle|1\rangle$. Then we apply $n + 1$ Hadamard gates, one for each qubit, in order to obtain the equal superposition of all basis states for the $|x\rangle$ register and the state $\frac{1}{\sqrt{2}}(|0\rangle - |1\rangle)$ for the ancillary qubit. Then we evaluate the oracle function, $|x\rangle|y\rangle \rightarrow |x\rangle|y \oplus f(x)\rangle$. As for the case of $n = 2$ qubits, this function evaluation kicks the sign back in front of the $|x\rangle$ register:

$$\frac{1}{\sqrt{2^n}} \sum_{x=0}^{2^n-1} |x\rangle \frac{1}{\sqrt{2}}(|0\rangle - |1\rangle)$$

$$\rightarrow \frac{1}{\sqrt{2^n}} \sum_{x=0}^{2^n-1} |x\rangle \frac{1}{\sqrt{2}}(|0 \oplus f(x)\rangle - |1 \oplus f(x)\rangle)$$

$$= \frac{1}{\sqrt{2^n}} \sum_{x=0}^{2^n-1} |x\rangle \frac{1}{\sqrt{2}}(-1)^{f(x)}(|0\rangle - |1\rangle)$$

$$= \frac{1}{\sqrt{2^n}} \sum_{x=0}^{2^n-1} (-1)^{f(x)}|x\rangle \frac{1}{\sqrt{2}}(|0\rangle - |1\rangle). \qquad (3.122)$$

However, we point out that, unlike the previous elementary example with $n = 2$ qubits, a single evaluation of the function $f(x)$ is not sufficient to find the marked item x_0. We must iterate the oracle query, which evaluates the function $f(x)$, several times. More precisely, we have to iterate several times the so-called Grover iteration G, defined as $G = DO$, where O denotes the oracle query and

$$D = H^{\otimes n}(-I + 2|0\rangle\langle 0|)H^{\otimes n}. \qquad (3.123)$$

The transformation in between the two n-qubit Hadamard gates in (3.123) is a conditional phase shift, that puts a phase shift of -1 in front of all the states of the computational basis, except for the state $|00\ldots0\rangle$. The quantum search algorithm is performed by repeatedly applying G until the state of the register $|x\rangle$ is such that a standard measurement gives the outcome x_0 with high probability.

3.10.3 *Geometric visualization*

A simple geometric visualization helps in understanding the number of times that Grover's iteration G has to be applied. To this end we can ignore the ancillary qubit, whose state is always factorized and never changes.

First of all, it is clear from Eq. (3.122) that the action of the oracle O on an n-qubit state vector $|x\rangle$ is given by

$$O \, : \, |x\rangle \rightarrow (-)^{f(x)}|x\rangle. \tag{3.124}$$

Therefore, we can write

$$O = I - 2|x_0\rangle\langle x_0| \equiv R_{|0\rangle}, \tag{3.125}$$

which corresponds to a reflection about the hyperplane perpendicular to $|x_0\rangle$. For example, if we consider a bidimensional space spanned by the vectors $\{|x_0\rangle, |x_0^\perp\rangle\}$ and a generic vector $|\psi\rangle = \alpha|x_0\rangle + \beta|x_0^\perp\rangle$, we have $O|\psi\rangle = -\alpha|x_0\rangle + \beta|x_0^\perp\rangle$. Thus, as appears clearly from Fig. 3.24, O induces a reflection of the vector $|\psi\rangle$ about the axis $|x_0^\perp\rangle$, that is, about the hyperplane perpendicular to $|x_0\rangle$.

Next we consider the transformation D. Note that

$$\left(H^{\otimes n}\right)^\dagger\left(-I + 2|0\rangle\langle 0|\right)H^{\otimes n} = -I + 2|S\rangle\langle S|, \tag{3.126}$$

where we have exploited the fact that $(H^{\otimes n})^\dagger = H^{\otimes n}$ and where $|S\rangle$ is the equal superposition state:

$$|S\rangle \equiv H^{\otimes n}|0\rangle = \frac{1}{\sqrt{2^n}}\sum_{x=0}^{2^n-1}|x\rangle. \tag{3.127}$$

Thus, we can write Eq. (3.123) as minus the reflection about the hyperplane orthogonal to $|S\rangle$:

$$D = -\left(I - 2|S\rangle\langle S|\right) \equiv -R_{|S\rangle}. \tag{3.128}$$

Therefore,

$$G = DO = -R_{|S\rangle}\,R_{|0\rangle}. \tag{3.129}$$

We now consider the two-dimensional plane spanned by $\{|x_0\rangle, |S\rangle\}$ and in this plane we draw the corresponding perpendicular unit vectors

$\{|x_0^{\perp}\rangle, |S^{\perp}\rangle\}$ (see Fig. 3.24). It is easy to prove[4] that

$$G = -R_{|S\rangle}R_{|0\rangle} = R_{|S^{\perp}\rangle}R_{|0\rangle}. \tag{3.130}$$

Thus, if θ denotes the angle between the vectors $|x_0^{\perp}\rangle$ and $|S\rangle$, G rotates a generic vector $|\psi\rangle$ in this plane by an angle 2θ (see Fig. 3.24). Since prior to the first oracle query the n-qubit state is the equal superposition state

$$|\psi_0\rangle \equiv |S\rangle = \sin\theta|x_0\rangle + \cos\theta|x_0^{\perp}\rangle, \tag{3.131}$$

and G rotates any vector in the plane spanned by $\{|x_0\rangle, |S\rangle\}$ by an angle 2θ, then, after j steps of Grover's iteration, the n-qubit state is given by

$$|\psi_j\rangle \equiv G^j|\psi_0\rangle = \sin\left((2j+1)\theta\right)|x_0\rangle + \cos\left((2j+1)\theta\right)|x_0^{\perp}\rangle. \tag{3.132}$$

We note that this state always belongs to the plane of Fig. 3.24. The process must stop after k steps, where k is such that $|\psi_k\rangle$ is very close to the marked state $|x_0\rangle$. This takes place when $|\sin(2k+1)\theta| \approx 1$. The smallest integer k that fulfils this condition is determined by the following relation:

$$(2k+1)\theta \approx \frac{\pi}{2}, \tag{3.133}$$

which implies

$$k = \text{round}\left(\frac{\pi}{4\theta} - \frac{1}{2}\right), \tag{3.134}$$

where round signifies the nearest integer. Since one starts from the equal superposition state,

$$\sin\theta = \langle x_0|\psi_0\rangle = \frac{1}{\sqrt{N}}, \tag{3.135}$$

and therefore, for large N, $\theta \approx \frac{1}{\sqrt{N}}$. Thus, the number of required iterations in Grover's algorithm is

$$k = \text{round}\left(\frac{\pi}{4}\sqrt{N} - \frac{1}{2}\right) = O(\sqrt{N}). \tag{3.136}$$

The algorithm ends with a standard measurement in the computational basis, giving outcome $x = \bar{x}$. Then we proceed as in any probabilistic algorithm: we check by means of the oracle if the solution is correct. This is the case if $f(\bar{x}) = 1$, so that $\bar{x} = x_0$. If this is not the case, that is, $f(\bar{x}) = 0$ implying $\bar{x} \neq x_0$, we repeat the quantum computation from the beginning. Therefore, unlike the special case with $N = 4$, the Grover

---

[4]In order to prove that $R_{|S\rangle} = -R_{|S^{\perp}\rangle}$, we consider a generic vector $|u\rangle = \mu|S\rangle + \nu|S^{\perp}\rangle$ residing in the plane spanned by $\{|S\rangle, |S^{\perp}\rangle\}$. We have $R_{|S\rangle}|u\rangle = -\mu|S\rangle + \nu|S^{\perp}\rangle$ while $R_{|S^{\perp}\rangle}|u\rangle = \mu|S\rangle - \nu|S^{\perp}\rangle = -R_{|S\rangle}|u\rangle$.

algorithm succeeds with a probability that is not equal to one. However, the success probability is very close to one (see exercise 3.18).

We note that the geometric interpretation of Grover's algorithm is consistent with the particular case $N = 4$, in which $\theta = \frac{\pi}{6}$, and therefore condition (3.133) is fulfilled after $k = 1$ step.

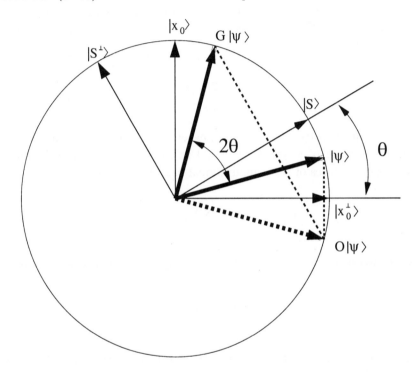

Fig. 3.24   Geometric visualization of Grover's iteration.

Exercise 3.18   Show that the probability that Grover's algorithm fails (*i.e.*, the measurement does not give the marked item x_0) drops like $1/N$.

Exercise 3.19   Estimate the number of elementary gates required to compute one step of Grover's iteration.

We point out that Grover's algorithm has been extended to the search for multiple items (also in the case in which the number of marked items is not known in advance) and many other situations. Unfortunately, it is also possible to prove that Grover's algorithm is optimal; that is, any other quantum algorithm for searching in an unstructured database would require at least $O(\sqrt{N})$ oracle queries.

3.11 The quantum Fourier transform

The *discrete* Fourier transform of a vector with complex components $\{f(0), f(1), \ldots, f(N-1)\}$ is a new complex vector $\{\tilde{f}(0), \tilde{f}(1), \ldots, \tilde{f}(N-1)\}$, defined as

$$\tilde{f}(k) = \frac{1}{\sqrt{N}} \sum_{j=0}^{N-1} e^{2\pi i \frac{jk}{N}} f(j). \tag{3.137}$$

The quantum Fourier transform does exactly the same. It is defined on a quantum register of n qubits ($N = 2^n$) as the unitary operator F whose action on the states of the computational basis is given by:

$$F(|j\rangle) = \frac{1}{\sqrt{2^n}} \sum_{k=0}^{2^n-1} e^{2\pi i \frac{jk}{2^n}} |k\rangle. \tag{3.138}$$

As a consequence, an arbitrary state $|\psi\rangle = \sum_j f(j)|j\rangle$ is transformed into

$$|\tilde{\psi}\rangle = F|\psi\rangle = \sum_{k=0}^{2^n-1} \tilde{f}(k)|k\rangle, \tag{3.139}$$

where the coefficients $\{\tilde{f}(k)\}$ are the discrete Fourier transform of the coefficients $\{f(j)\}$, according to the relation (3.137).

Now we construct the quantum circuit for computing the quantum Fourier transform. It is useful to introduce the following notations for the binary representation of j:

$$j = j_{n-1} j_{n-2} \cdots j_0 = j_{n-1} 2^{n-1} + j_{n-2} 2^{n-2} + \cdots + j_0 2^0, \tag{3.140}$$

and for a binary fraction:

$$0.j_l j_{l+1} \cdots j_m = \frac{1}{2} j_l + \frac{1}{4} j_{l+1} + \cdots + \frac{1}{2^{m-l+1}} j_m. \tag{3.141}$$

In a few simple steps we obtain the *product representation* of the Fourier transform:

$$
\begin{aligned}
F(|j\rangle) &= \frac{1}{\sqrt{2^n}} \sum_{k=0}^{2^n-1} \exp\left(\frac{2\pi i j k}{2^n}\right) |k\rangle \\
&= \frac{1}{\sqrt{2^n}} \sum_{k_{n-1}=0}^{1} \cdots \sum_{k_0=0}^{1} \exp\left(2\pi i j \sum_{l=1}^{n} \frac{k_{n-l}}{2^l}\right) |k_{n-1} \ldots k_0\rangle \\
&= \frac{1}{\sqrt{2^n}} \sum_{k_{n-1}=0}^{1} \cdots \sum_{k_0=0}^{1} \otimes_{l=1}^{n} \exp\left(2\pi i j \frac{k_{n-l}}{2^l}\right) |k_{n-l}\rangle \\
&= \frac{1}{\sqrt{2^n}} \otimes_{l=1}^{n} \left[\sum_{k_{n-l}=0}^{1} \exp\left(2\pi i j \frac{k_{n-l}}{2^l}\right) |k_{n-l}\rangle\right] \\
&= \frac{1}{\sqrt{2^n}} \otimes_{l=1}^{n} \left[|0\rangle + \exp\left(2\pi i j \frac{1}{2^l}\right) |1\rangle\right] \\
&= \frac{1}{\sqrt{2^n}} \left(|0\rangle + e^{2\pi i 0.j_0}|1\rangle\right) \left(|0\rangle + e^{2\pi i 0.j_1 j_0}|1\rangle\right) \cdots \\
&\qquad\qquad \cdots \left(|0\rangle + e^{2\pi i 0.j_{n-1}j_{n-2}\cdots j_0}|1\rangle\right). \quad (3.142)
\end{aligned}
$$

It is interesting to note that this product representation is factorized; this shows that the corresponding quantum state is *not entangled*.

The product representation (3.142) makes it easy to construct a quantum circuit that computes the quantum Fourier transform *efficiently*. We show such a circuit in Fig. 3.25, where R_k denotes the operator

$$
R_k = \begin{bmatrix} 1 & 0 \\ 0 & \exp\left(\frac{2\pi i}{2^k}\right) \end{bmatrix}. \quad (3.143)
$$

We consider the action of the circuit on a state $|j\rangle = |j_{n-1}j_{n-2}\ldots j_0\rangle$ of the computational basis (for a generic state $|\psi\rangle = \sum_j c_j |j\rangle$ it is sufficient to remember that the Fourier transform is a linear operator). The first Hadamard gate acts on the most significant qubit and generates the state

$$
\frac{1}{\sqrt{2}} \left(|0\rangle + e^{2\pi i 0.j_{n-1}}|1\rangle\right) |j_{n-2} \cdots j_0\rangle. \quad (3.144)
$$

The subsequent controlled phase rotations, controlled-R_2 to controlled-R_n, add phases from $\pi/2$ to $\pi/2^{n-1}$ if the corresponding control qubit is set to one. After these $n-1$ two-qubit gates, the quantum-computer wave function is left in the state

$$
\frac{1}{\sqrt{2}} \left(|0\rangle + e^{2\pi i 0.j_{n-1}j_{n-2}\cdots j_0}|1\rangle\right) |j_{n-2} \cdots j_0\rangle. \quad (3.145)
$$

A similar procedure is then repeated for the other qubits and therefore the
quantum circuit in Fig. 3.25 generates the output

$$\frac{1}{\sqrt{2^n}} \left(|0\rangle + e^{2\pi i \, 0.j_{n-1}j_{n-2}\cdots j_0}|1\rangle\right) \left(|0\rangle + e^{2\pi i \, 0.j_{n-2}\cdots j_0}|1\rangle\right) \cdots$$
$$\cdots \left(|0\rangle + e^{2\pi i \, 0.j_0}|1\rangle\right). \qquad (3.146)$$

This state coincides with the product representation (3.142), except for
the fact that the order of the qubits is reversed. The correct order can
be obtained by means of $O(n)$ SWAP gates (see Fig. 3.4); alternatively,
one can simply relabel qubits. The circuit of Fig. 3.25 shows that the
discrete Fourier transform of a complex vector of size $N = 2^n$ can be
implemented efficiently on a quantum register of n qubits using n Hadamard
and $n(n-1)/2$ controlled phase-shift gates. Therefore, the computation of
the quantum Fourier transform requires $O(n^2)$ elementary quantum gates,
whereas the most efficient classical algorithm, the *fast Fourier transform*,
computes the discrete Fourier transform in $O(2^n n)$ elementary operations.
    However, we emphasize that we cannot really talk about an exponential
speed up in the computation of the Fourier transform, since a generic state
$|\psi\rangle = \sum_j f(j)|j\rangle$ cannot be prepared efficiently (see the discussion at the
end of Sec. 3.5) and the final state $|\tilde{\psi}\rangle = \sum_k \tilde{f}(k)|k\rangle$ is not readily acces-
sible. Indeed, a standard measurement simply gives an outcome $\overline{k}$ with
probability $\left|\tilde{f}(\overline{k})\right|^2$. The problem is that the quantum Fourier transform is
performed on the amplitudes of the wave function, which are *not directly
accessible*. They can only be reconstructed with finite accuracy after many
runs (each run computes the Fourier transform of the state $|\psi\rangle$ and ends
up with a standard projective measurement). If N denotes the number of
runs, we can estimate $\tilde{f}(k)$ as N_k/N, where N_k is the number of times that
the measurement gives outcome k. We already encountered this problem in
relation to function evaluation in Sec. 3.7 and it is actually a typical difficulty of quantum computation. Quantum algorithms find a way to *extract*
efficiently useful information from the quantum-computer wave function.
As we shall see in the following sections, the quantum Fourier transform is
a key ingredient of exponentially efficient quantum algorithms.

Exercise 3.20   Estimate the effect of unitary errors on the stability of
the quantum Fourier transform.

Exercise 3.21   Construct a quantum circuit to compute the inverse

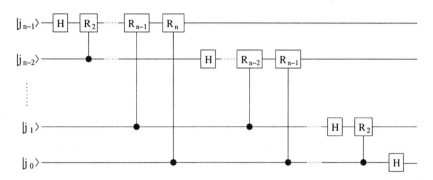

Fig. 3.25   A circuit implementing the quantum Fourier transform. The output is given by Eq. (3.146).

Fourier transform

$$F^{-1}(|j\rangle) = \frac{1}{\sqrt{2^n}} \sum_{k=0}^{2^n-1} e^{-2\pi i \frac{jk}{2^n}} |k\rangle. \qquad (3.147)$$

3.12   Quantum phase estimation

As a first application of the quantum Fourier transform we consider the following problem: a unitary operator U has an eigenvector $|u\rangle$ with eigenvalue $e^{i\phi}$ ($0 \le \phi < 2\pi$). Assume that we are able to prepare the state $|u\rangle$ and there is a black box routine capable of performing the operations controlled-U^{2^j}, where j is a non-negative integer. We wish to obtain the best n-bit estimate of ϕ.

The quantum circuit solving this problem is shown in Fig. 3.26. The first register contains n qubits, n depending on the required accuracy for ϕ. The second register contains the m qubits necessary to store $|u\rangle$. The action of the gate $C - U^{2^j}$ on a state $\frac{1}{\sqrt{2}}(|0\rangle + |1\rangle)|u\rangle$ is given by

$$
\begin{aligned}
C - U^{2^j} \tfrac{1}{\sqrt{2}}(|0\rangle + |1\rangle)|u\rangle &= \tfrac{1}{\sqrt{2}}(|0\rangle|u\rangle + |1\rangle\, U^{2^j}|u\rangle) \\
&= \tfrac{1}{\sqrt{2}}(|0\rangle|u\rangle + |1\rangle e^{i2^j\phi}|u\rangle) \\
&= \tfrac{1}{\sqrt{2}}(|0\rangle + e^{i2^j\phi}|1\rangle)|u\rangle. \qquad (3.148)
\end{aligned}
$$

Taking into account this result, it is easy to check that the output of the circuit (3.26) is given by

$$\frac{1}{\sqrt{2^n}}\left(|0\rangle + e^{i2^{n-1}\phi}|1\rangle\right)\cdots\left(|0\rangle + e^{i2\phi}|1\rangle\right)\left(|0\rangle + e^{i\phi}|1\rangle\right)|u\rangle$$

$$= \frac{1}{\sqrt{2^n}}\sum_{y=0}^{2^n-1} e^{i\phi y}|y\rangle|u\rangle. \tag{3.149}$$

As in Deutsch's and Grover's algorithms, the key point is that the quantum register that stores $|u\rangle$ is prepared in an eigenstate of the operators U, U^2, U^4, ... As a consequence, the state of this register never changes and the phase factors $e^{i\phi}$, $e^{2i\phi}$, $e^{4i\phi}$, ... are propagated backwards in the control register.

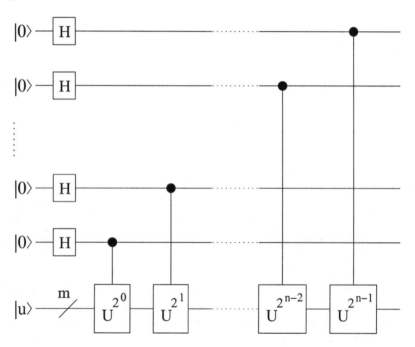

Fig. 3.26   A quantum circuit used to obtain the state (3.149) in the phase estimation problem. The wire with a dash represents a set of m qubits.

From now on we consider only the control register and show that its quantum Fourier transform gives the best n-bit estimate of ϕ with high probability. It is convenient to write

$$\phi = 2\pi\left(\frac{a}{2^n} + \delta\right), \tag{3.150}$$

where $a = a_{n-1}a_{n-2} \cdots a_1 a_0$ (in binary notation), $2\pi a/2^n$ is the best n-bit approximation of ϕ and therefore $0 \le |\delta| \le 1/2^{n+1}$. One can check that the inverse quantum Fourier transform (defined by Eq. 3.147) of the first register, applied to the state (3.149), gives

$$\frac{1}{2^n} \sum_{x=0}^{2^n-1} \sum_{y=0}^{2^n-1} e^{2\pi i(a-x)y/2^n} e^{2\pi i \delta y} |x\rangle . \tag{3.151}$$

We now perform a standard measurement of the first ($|x\rangle$) register. If $\delta = 0$, the wave vector (3.151) reduces to $|a\rangle$. Therefore, in this case a standard measurement of the first register gives outcome a with certainty and the phase ϕ is determined exactly. In the general case $\delta \ne 0$, the best n-bit estimate of ϕ is given by a and is obtained from a standard measurement of the first register with probability $p_a = |c_a|^2$. Here c_a denotes the projection of the wave function (3.151) over the state $|a\rangle$ and is given by

$$c_a = \frac{1}{2^n} \sum_{y=0}^{2^n-1} \left(e^{2\pi i \delta}\right)^y = \frac{1}{2^n} \sum_{y=0}^{2^n-1} \alpha^y \qquad (\alpha \equiv e^{2\pi i \delta}); \tag{3.152}$$

this finite geometric series can be added, giving

$$c_a = \frac{1}{2^n} \left[\frac{1-\alpha^{2^n}}{1-\alpha} \right]. \tag{3.153}$$

Since for any $z \in [0, \frac{1}{2}]$ we have $2z \le \sin(\pi z) \le \pi z$, we obtain

$$\left| 1 - e^{2\pi i \delta 2^n} \right| = 2|\sin(\pi \delta 2^n)| \ge 4|\delta| 2^n,$$

$$\left| 1 - e^{2\pi i \delta} \right| = 2|\sin(\pi \delta)| \le 2\pi |\delta|. \tag{3.154}$$

We can insert these two inequalities into (3.153) and obtain

$$|c_a|^2 \ge \frac{4}{\pi^2} \approx 0.405 . \tag{3.155}$$

Therefore, the best n-bit estimate of ϕ is obtained with high probability $|c_a|^2$.

We point out that it is possible to obtain the best l-bit approximation of the phase ϕ with probability arbitrarily close to 1, provided that the number n of qubits is large enough. More precisely, the best l-bit approximation of ϕ is obtained with probability $> 1 - \epsilon$ if the circuit of Fig. 3.26 contains

$n = l + O(\log(1/\epsilon))$ qubits in the first register (see Cleve *et al.*, 1998). Thus, by increasing the number n we raise not only the accuracy of our phase estimation but also the probability that our algorithm succeeds.

In summary, the quantum-phase estimation algorithm exploits the ability of the inverse quantum Fourier transform to perform the transformation from the state (3.149) to the state (3.151). As we have seen above, this latter state, when measured, gives with high probability a good estimate of the phase ϕ. We stress that this algorithm is exponentially efficient with respect to any known classical algorithm solving the phase estimation problem, provided that the unitary operator U can be decomposed efficiently into elementary gates on a quantum computer (that is, $U|u\rangle$ can be computed in a number of elementary quantum gates polynomial in the number m of qubits necessary to store $|u\rangle$).

3.13 * Finding eigenvalues and eigenvectors

In this section, we describe a quantum algorithm that computes eigenvalues and eigenvectors of a given unitary operator U. To be concrete, we consider the case in which

$$U(\bar{t}) = \exp(-iH\bar{t}/\hbar) \tag{3.156}$$

is the evolution operator up to time $\bar{t}$, associated with a time-independent Hamiltonian H. The corresponding Schrödinger equation is

$$i\hbar \frac{\partial}{\partial t} \psi(x,t) = H(x)\psi(x,t), \tag{3.157}$$

where, for the sake of simplicity, we have considered the one-dimensional case (the extension of what follows to higher dimensions is trivial). We note that an eigenvector $\phi_\alpha(x)$ of H with eigenvalue E_α (*i.e.*, $H\phi_\alpha(x) = E_\alpha\phi_\alpha(x)$) is also an eigenvector of $U(\bar{t})$ with eigenvalue $e^{-iE_\alpha\bar{t}/\hbar}$.

The eigenvalues and eigenvectors of the Hamiltonian operator H can be computed using a classical computer as follows. We evolve some initial state

$$\psi_0(x) \equiv \psi(x,t=0), \tag{3.158}$$

obtaining $\psi(x,\bar{t}) = U(x,\bar{t})\psi_0(x)$. If we expand $\psi_0(x)$ over the basis of eigenfunctions of H,

$$\psi_0(x) = \sum_\alpha a_\alpha \phi_\alpha(x), \qquad (3.159)$$

we can write

$$\psi(x,t) = \sum_\alpha a_\alpha e^{-i\omega_\alpha t} \phi_\alpha(x) \qquad (\omega_\alpha \equiv E_\alpha/\hbar). \qquad (3.160)$$

Then we compute the Fourier transform

$$\tilde{\psi}(x_0,\omega) = F[\psi(x_0,t)], \qquad (3.161)$$

for a given $x = x_0$. It is evident from (3.160) that the Fourier transform $\tilde{\psi}$ exhibits peaks corresponding to the frequencies of motion ω_α. A given peak is resolved if the time evolution is computed up to time $\bar{t}$ much longer than the inverse of the frequency of interest. If the Fourier transform $\tilde{\psi}$ is repeated for different x_0 values, one obtains the eigenfunctions $\phi_\alpha(x)$ from the relative amplitudes of the peaks corresponding to the frequencies ω_α. This latter point is evident from the expansion (3.160), which implies

$$\frac{\tilde{\psi}(x_1,\omega_\alpha)}{\tilde{\psi}(x_2,\omega_\alpha)} = \frac{\phi_\alpha(x_1)}{\phi_\alpha(x_2)}. \qquad (3.162)$$

Let us now repeat the same procedure on a quantum computer. The wave function is coded inside some interval $x \in [-L, L]$ of interest for the physical motion, using a grid of 2^n points separated by an interval $\Delta x = 2L/(2^n - 1)$:

$$|\psi\rangle = \sum_{i=0}^{2^n-1} \psi(i) |i\rangle, \qquad (3.163)$$

where $\psi(i) = \psi(-L + i\Delta x)$. We note that the quantum computer has an exponential advantage in memory requirements, since one needs n qubits to store the 2^n complex numbers of the wave function (3.163). The evolution in a time step Δt is given by the unitary operator

$$U = \exp(-i H \Delta t/\hbar), \qquad (3.164)$$

which, for a large class of physically significant Hamiltonians, can be simulated efficiently on a quantum computer (see Lloyd, 1996). The appropriate circuit for computing the eigenvalues and eigenvectors of the operator U is that introduced in Fig. 3.26 for the phase estimation problem. The second (target) register (n qubits) is used to store the wave function, initially prepared in the state $|\psi_0\rangle = \sum_i \psi_0(i)|i\rangle$. Such a preparation in special cases can be performed efficiently, for example when the wave function is localized at a point, $|\psi_0\rangle = |\bar{i}\rangle$. We note that this choice is good enough to search for all eigenstates $|\phi_\alpha\rangle = \sum_i \phi_\alpha(i)|i\rangle$ (and corresponding eigenvalues) having non-zero projections over $|\psi_0\rangle$, namely, a non-zero component $\phi_\alpha(\bar{i})$. The control register (l qubits) is prepared in the uniform superposition state $\sum_j |j\rangle/\sqrt{2^l}$ and is necessary to simultaneously store the wave function at different times $0, \Delta t, 2\Delta t, 3\Delta t, \ldots, 2^{l-1}\Delta t$. After execution of the gates controlled-U^{2^0}, controlled-U^{2^1}, $\ldots$, controlled-$U^{2^{l-1}}$ (see Fig. 3.26) the state of the quantum computer is given by

$$|\Psi\rangle = \frac{1}{\sqrt{2^l}} \sum_{j=0}^{2^l-1} |j\rangle\, U^j\, |\psi_0\rangle\,, \qquad (3.165)$$

where we have

$$|\psi(j\Delta t)\rangle = U^j\, |\psi_0\rangle\,. \qquad (3.166)$$

Using the expansion (3.159), we can write

$$|\Psi\rangle = \frac{1}{\sqrt{2^l}} \sum_j |j\rangle \sum_{\alpha=0}^{2^n-1} a_\alpha\, e^{-i\omega_\alpha\, j\Delta t}\, |\phi_\alpha\rangle\,. \qquad (3.167)$$

It can be shown that the Fourier transform of the state $|\Psi\rangle$ with respect to the $|j\rangle$ register has peaks corresponding to the frequencies ω_α. A sufficiently large number of computer runs followed by a standard measurement of the $|j\rangle$ register would allow one to obtain the frequency spectrum of the Hamiltonian operator. It can be checked that, after any such measurement, the other quantum register collapses onto an eigenstate $|\phi_{\bar{\alpha}}\rangle$. The eigenstates can in principle be reconstructed by means of standard measurements of this register.

The limitation of the method is that each computer run singles out different eigenvalues ω_α, with probability $|a_\alpha|^2$. With respect to the classical algorithm described above, there is an exponential advantage if the desired eigenvalues and eigenvectors are obtained after a polynomial number of trials. This is, for instance, the case if the initial state $|\psi_0\rangle$ has a not exponentially small projection over the desired eigenstate, typically the ground state of a complex system (see Abrams and Lloyd, 1999).

3.14 Period finding and Shor's algorithm

The most spectacular discovery of quantum computation, Shor's algorithm, efficiently solves the *prime factorization* problem: given a composite odd positive integer N, find its prime factors. This is a central problem in computer science and it is conjectured, though not proven, that in classical computation this problem is in the class **NP** but not **P**: given the factors, it is easy to check if they solve the problem, but it is hard to find them. There are cryptographic systems (such as RSA, see Chap. 4) that are extensively used today and that are based on this conjecture. Indeed, in spite of the efforts over many centuries to find a polynomial time factoring algorithm, at present the best classical algorithm, the number field sieve, requires $\exp(O(n^{1/3}(\log n)^{2/3}))$ operations; that is, it is super-polynomial in the input size $n = \log N$. Shor discovered a quantum algorithm that accomplishes the same task in $O(n^2 \log n \log \log n)$ elementary gates. Therefore, this algorithm is polynomial in the input size and there is an exponential speed up with respect to any known classical algorithm.

The problem can be reduced to the problem of finding the period of the function $f(x) = a^x \bmod N$, where N is the number to be factorized and $a < N$ is chosen randomly. The steps to operate this reduction can be performed efficiently on a classical computer and we do not discuss them here (see Shor, 1997).

In the following, we consider the problem of finding the period r of a function $f(x)$, with $f(x + r) = f(x)$. To simplify the discussion, we shall consider the particular case in which r exactly divides the number of points $N = 2^n$ over which the function $f(x)$ is evaluated, that is $N/r = m$, with m integer. The general case adds some complications but does not change the general ideas discussed below. We need two registers, the first is prepared in the equal superposition state and the second stores the function $f(x)$, building the total state

$$\frac{1}{\sqrt{2^n}} \sum_{x=0}^{2^n-1} |x\rangle \, |f(x)\rangle \, . \tag{3.168}$$

Note that the modular exponentiation function required in Shor's algorithm can be computed efficiently on a quantum as well as on a classical computer (for the quantum case, see Barenco *et al.*, 1995; Miquel *et al.*, 1996). However, the *quantum parallelism* allows the function $f(x)$ to be computed for all x in a single run. We also note that after function evaluation the two registers are *entangled*. Unfortunately, we cannot gain direct access to all values $f(x)$. On the contrary, after a measurement of the second register, it collapses onto a particular state $|f(x_0)\rangle$. Thus, the quantum-computer wave function becomes

$$\frac{1}{\sqrt{m}} \sum_{j=0}^{m-1} |x_0 + jr\rangle\, |f(x_0)\rangle \qquad (0 \le x_0 < r - 1), \qquad (3.169)$$

where $m = N/r$ is the number of x values such that $f(x) = f(x_0)$. Since r is the period of $f(x)$, we have $f(x_0) = f(x_0 + r) = f(x_0 + 2r) = \cdots = f(x_0 + (m-1)r)$. We can now neglect the second register, which is factorized and does not concern us any more. It can be checked that a quantum Fourier transform of the first register gives the state

$$\frac{1}{\sqrt{r}} \sum_{k=0}^{r-1} \exp\left(\frac{2\pi i\, x_0\, k}{r}\right) \left|k\, \frac{N}{r}\right\rangle . \qquad (3.170)$$

Therefore, *quantum interference* has selected a few specific frequencies. Actually, a quantum measurement of the wave function (3.170) will give one of the r outcomes kN/r $(k = 0, 1, \ldots, r - 1)$ with equal probability. Thus, if we denote by c the measured value, we have $c/N = \lambda/r$, with λ an unknown integer. If λ and r have no common factors, then we reduce c/N to an irreducible fraction and thus obtain both λ and r. Number theory tells us that this takes place with probability at least $1/\log\log r$. Otherwise, the algorithm fails and must be repeated. It can be shown that the algorithm succeeds with probability arbitrarily close to one after a number of runs $O(\log\log r)$, see Shor (1997).

As a simple example, let us attempt to find the period of the function $f(x) = \frac{1}{2}(\cos(\pi x) + 1)$, assuming that x is loaded in a 3-qubit register, that is, $N = 2^3 = 8$. The function $f(x)$ may be equal to 0 or 1 and therefore can be stored in a single-qubit register. Unknown to us, the function has a period $r = 2$, since $f(x) = 1$ for even x and $f(x) = 0$ for odd x. We wish to determine this period using the period-finding algorithm. For this purpose, we start from the fiducial state $|000\rangle|0\rangle$ and build the state (3.168), which reads

$$\frac{1}{\sqrt{8}} \big(|0\rangle\,|f(0)\rangle + |1\rangle\,|f(1)\rangle + |2\rangle\,|f(2)\rangle + |3\rangle\,|f(3)\rangle$$
$$+ |4\rangle\,|f(4)\rangle + |5\rangle\,|f(5)\rangle + |6\rangle\,|f(6)\rangle + |7\rangle\,|f(7)\rangle\big). \qquad (3.171)$$

Then we measure the second register obtaining, for example, the outcome 0. Thus, the total wave function collapses onto the state

$$\tfrac{1}{2}\left(|1\rangle + |3\rangle + |5\rangle + |7\rangle\right)|0\rangle, \tag{3.172}$$

where $x = 1, 3, 5, 7$ are the values such that $f(x) = 0$. From now on, we neglect the second register, which is no longer used. Next, we apply the quantum Fourier transform (described in Sec. 3.11) to the first register, leading to the following transformation:

$$|x\rangle \rightarrow \tfrac{1}{\sqrt{8}} \sum_{k=0}^{7} e^{2\pi i k x/8}|k\rangle. \tag{3.173}$$

As a result, we obtain the wave function

$$\begin{aligned}
|\psi\rangle &= \tfrac{1}{2\sqrt{8}}\left(|0\rangle + e^{i\pi/4}|1\rangle + e^{i2\pi/4}|2\rangle + \cdots + e^{i7\pi/4}|7\rangle\right) \\
&+ \tfrac{1}{2\sqrt{8}}\left(|0\rangle + e^{i3\pi/4}|1\rangle + e^{i6\pi/4}|2\rangle + \cdots + e^{i21\pi/4}|7\rangle\right) \\
&+ \tfrac{1}{2\sqrt{8}}\left(|0\rangle + e^{i5\pi/4}|1\rangle + e^{i10\pi/4}|2\rangle + \cdots + e^{i35\pi/4}|7\rangle\right) \\
&+ \tfrac{1}{2\sqrt{8}}\left(|0\rangle + e^{i7\pi/4}|1\rangle + e^{i14\pi/4}|2\rangle + \cdots + e^{i49\pi/4}|7\rangle\right). \tag{3.174}
\end{aligned}$$

It is easy to check that the complex amplitudes in front of the states $|1\rangle$, $|2\rangle$, $|3\rangle$, $|5\rangle$, $|6\rangle$ and $|7\rangle$ cancel each other. The interference is constructive only for the terms in front of the states $|0\rangle$ and $|4\rangle$. Thus, we can write the wave function $|\psi\rangle$ as follows:

$$|\psi\rangle = \tfrac{1}{\sqrt{2}}\left(|0\rangle - |4\rangle\right). \tag{3.175}$$

This expression is in agreement with the general formula (3.170). Finally, from the measurement of the first register we obtain outcomes 0 or 4 with equal probability. In the first case, we cannot find the period r of the function $f(x)$ and the algorithm must be repeated. In the latter case, we write $c/N = \lambda/r$, with $c = 4$ the measured value. Eventually, we reduce $c/N = 4/8$ to an irreducible fraction $c/N = \lambda/r = 1/2$, thus obtaining the period $r = 2$.

Turning to the power of quantum computation, the following question naturally arises: Why is the speedup of quantum computation (with respect to any known classical algorithm) exponential for the factoring problem and only quadratic for the searching problem? Let us give an intuitive argument to help understanding this fact. The searching problem is a typical structureless problem. Indeed, we search for an item in a database without any structure. Fortunately, quantum mechanics helps this search giving a significant quadratic speedup. Unfortunately, it turns out that this gain is is the maximum possible. On the other hand, Shor's algorithm exploits a structure hidden in the factoring problem. This structure allows the reduction of the integer-factoring problem to the problem of finding the period of a particular function. The natural way to find the period of a function is to compute its Fourier transform and, as we have seen, the Fourier transform can be implemented efficiently on a quantum computer. This being said, we do not know the answer to the following fundamental question: What class of problems can be simulated efficiently on a quantum computer? Are there other problems for which the quantum computer gives an exponential gain, beyond those based on the quantum Fourier transform?

3.15 Quantum computation of dynamical systems

In this section, we show that a quantum computer would be useful to simulate the dynamical evolution of quantum systems. The simulation of quantum many-body problems on a classical computer is a difficult task as the size of the Hilbert space grows exponentially with the number of particles. For instance, if we wish to simulate a chain of n spin-$\frac{1}{2}$ particles, the size of the Hilbert space is 2^n. Namely, the state of this system is determined by 2^n complex numbers. As observed by Feynman (1982), the growth in memory requirement is only linear on a quantum computer, which is itself a many-body quantum system. For example, to simulate n spin-$\frac{1}{2}$ particles we only need n qubits. Therefore, a quantum computer operating with only a few tens of qubits can outperform a classical computer. Of course, this is only true if we can find an efficient quantum algorithm and if we can efficiently extract useful information from the quantum computer. We shall discuss this point for a specific quantum algorithm later in this section.

3.15.1 *Quantum simulation of the Schrödinger equation*

To be concrete, let us consider the quantum-mechanical motion of a particle in one dimension (the extension to higher dimensions is straightforward). It is governed by the Schrödinger equation

$$i\hbar \frac{d}{dt} \psi(x,t) = H \psi(x,t), \tag{3.176}$$

where the Hamiltonian H is given by

$$H = H_0 + V(x) = -\frac{\hbar^2}{2m}\frac{d^2}{dx^2} + V(x).$$ (3.177)

The Hamiltonian $H_0 = -(\hbar^2/2m)\,d^2/dx^2$ governs the free motion of the particle, while $V(x)$ is a one-dimensional potential. To solve Eq. (3.176) on a quantum computer with finite resources (a finite number of qubits and a finite sequence of quantum gates), we must first of all discretize the continuous variables x and t. If the motion essentially takes place inside a finite region, say $-d \leq x \leq d$, we decompose this region into 2^n intervals of length $\Delta = 2d/2^n$ and represent these intervals by means of the Hilbert space of an n-qubit quantum register (this means that the discretization step drops exponentially with the number of qubits). Hence, the wave function $|\psi(t)\rangle$ is approximated as follows:

$$|\tilde{\psi}(t)\rangle = \sum_{i=0}^{2^n-1} c_i(t)\,|i\rangle = \frac{1}{\mathcal{N}}\sum_{i=0}^{2^n-1}\psi(x_i,t)\,|i\rangle,$$ (3.178)

where

$$x_i \equiv -d + \left(i + \tfrac{1}{2}\right)\Delta,$$ (3.179)

$|i\rangle = |i_{n-1}\rangle \otimes |i_{n-2}\rangle \otimes \cdots \otimes |i_0\rangle$ is a state of the computational basis of the n-qubit quantum register and

$$\mathcal{N} \equiv \sqrt{\sum_{i=0}^{2^n-1}|\psi(x_i,t)|^2}$$ (3.180)

is a factor that ensures correct normalization of the wave function. It is intuitive that $|\tilde{\psi}\rangle$ provides a good approximation to $|\psi\rangle$ when the discretization step Δ is much smaller than the shortest length scale relevant for the motion of the system.

As we have seen in Sec. 2.4, the Schrödinger equation (3.176) may be integrated formally by propagating the initial wave function $\psi(x,0)$ for each time-step ϵ as follows:

$$\psi(x,t+\epsilon) = e^{-\frac{i}{\hbar}[H_0+V(x)]\epsilon}\,\psi(x,t).$$ (3.181)

If the time-step ϵ is small enough, so that terms of order ϵ^2 may be neglected, it is possible to write

$$e^{-\frac{i}{\hbar}[H_0+V(x)]\epsilon} \approx e^{-\frac{i}{\hbar}H_0\,\epsilon}e^{-\frac{i}{\hbar}V(x)\,\epsilon}.$$ (3.182)

Note that this equation, known as the Trotter decomposition, is only exact up to terms of order ϵ^2 since the operators H_0 and V do not commute. The operator on the right-hand side of Eq. (3.182) is still unitary, simpler than that on the left-hand side, and, in many interesting physical problems, can be efficiently implemented on a quantum computer. We take advantage of the fact that, as we have seen in Sec. 3.11, the Fourier transform can be efficiently preformed by a quantum computer. We call k the variable conjugate to x, that is, $-i(d/dx) = F^{-1}kF$, where F is the Fourier transform. Therefore, we can write the first operator in the right-hand side of (3.182) as

$$e^{-\frac{i}{\hbar}H_0\,\epsilon} \;=\; F^{-1}\,e^{+\frac{i}{\hbar}\left(\frac{\hbar^2 k^2}{2m}\right)\epsilon}\,F\,.\qquad(3.183)$$

In this expression, we pass, by means of the Fourier transform F, from the x-representation to the k-representation, in which this operator is diagonal. Then, using the inverse Fourier transform F^{-1}, we return to the x-representation, in which the operator $\exp(-iV(x)\epsilon/\hbar)$ is diagonal. The wave function $\psi(x,t)$ at time $t = l\epsilon$ is obtained from the initial wave function $\psi(x,0)$ by applying l times the unitary operator

$$F^{-1}\,e^{+\frac{i}{\hbar}\left(\frac{\hbar^2 k^2}{2m}\right)\epsilon}\,F\,e^{-\frac{i}{\hbar}V(x)\,\epsilon}\,.\qquad(3.184)$$

Therefore, simulation of the Schrödinger equation is now reduced to the implementation of the Fourier transform plus diagonal operators of the form

$$|x\rangle \;\to\; e^{icf(x)}\,|x\rangle\,,\qquad(3.185)$$

where c is some real constant. Note that an operator of the form (3.185) appears both in the computation of $\exp(-iV(x)\epsilon/\hbar)$ and of $\exp(-iH_0\epsilon/\hbar)$, when this latter operator is written in the k-representation. The construction of the Fourier transform was discussed in Sec. 3.11 and requires $O(n^2)$ elementary quantum gates (Hadamard and controlled phase-shift gates). The quantum computation of (3.185) is possible, using an ancillary quantum register $|y\rangle_a$, by means of the following steps:

$$|0\rangle_a \otimes |x\rangle \;\to\; |f(x)\rangle_a \otimes |x\rangle \;\to\; e^{icf(x)}\,|f(x)\rangle_a \otimes |x\rangle$$
$$\to\; e^{icf(x)}\,|0\rangle_a \otimes |x\rangle \;=\; |0\rangle_a \otimes e^{icf(x)}\,|x\rangle\,.\qquad(3.186)$$

The first step is a standard function evaluation and, as we have discussed in Sec. 3.7, may be implemented by means of $O(2^n)$ generalized C^n-NOT gates. Of course, more efficient implementations (polynomial in n) are possible when the function $f(x)$ has some structure. This is the case for the potentials $V(x)$ usually considered in quantum-mechanical problems. The second step in (3.186) is the transformation $|y\rangle_a \to e^{icy}|y\rangle_a$ and can be performed in m single-qubit phase-shift gates, m being the number of qubits in the ancillary register. Indeed, we may write the binary decomposition of an integer $y \in [0, 2^m - 1]$ as $y = \sum_{j=0}^{m-1} y_j 2^j$, with $y_j \in \{0, 1\}$. Therefore,

$$\exp(iy) = \exp\left(\sum_{j=0}^{m-1} icy_j 2^j\right) = \prod_{j=0}^{m-1} \exp(icy_j 2^j), \qquad (3.187)$$

which is the product of m single-qubit gates, each acting non-trivially (differently from the identity) only on a single qubit. The j-th gate operates the transformation $|y_j\rangle_a \to \exp(icy_j 2^j)|y_j\rangle_a$, with $|y_j\rangle_a \in \{|0\rangle, |1\rangle\}$ vectors of the computational basis for the j-th ancillary qubit. The matrix representation of this phase shift gate is given by

$$R_z(c\,2^j) = \begin{bmatrix} 1 & 0 \\ 0 & \exp(ic\,2^j) \end{bmatrix}. \qquad (3.188)$$

The third step in (3.186) is just the reverse of the first and may be implemented by the same array of gates as the first but applied in the reverse order. After this step the ancillary qubits are returned to their standard configuration $|0\rangle_a$ and it is therefore possible to use the same ancillary qubits for every time-step.

Note that the number of ancillary qubits m determines the resolution in the computation of the diagonal operator (3.185). Indeed, the function $f(x)$ appearing in (3.185) is discretized and can take 2^m different values. It turns out that standard quantum-mechanical problems (the unidimensional harmonic and anharmonic oscillator, transmission and reflection of a Gaussian wave packet by various barriers, *etc.*) can be simulated with sufficient resolution using only a small number of qubits ($n \approx 10$) to discretize the coordinate x and a small number of ancillary qubits ($m \approx 10$), see Strini (2002).

In the following subsections, we shall discuss two interesting dynamical systems, the baker's map and the sawtooth map, that can be simulated on a quantum computer without ancillary qubits. In these models interesting physical phenomena can already be studied with 3–10 qubits and a few tens–hundreds of quantum gates.[5]

---

[5]It is interesting to note that, as we have seen in Sec. 3.10, Grover's search algorithm

3.15.2 * The quantum baker's map

In this subsection, we show that the quantum baker's map can be simulated on a quantum computer in a particularly simple and efficient way. The quantum algorithm for the quantum baker's map, proposed by Schack, computes the dynamical evolution of this system exponentially faster than any known classical computation.

The classical baker's transformation maps the unit square $0 \leq q, p < 1$ onto itself according to

$$(q,p) \rightarrow (\bar{q},\bar{p}) = \begin{cases} (2q, \frac{1}{2}p), & \text{if } 0 \leq q \leq \frac{1}{2}, \\ (2q - 1, \frac{1}{2}p + \frac{1}{2}), & \text{if } \frac{1}{2} < q < 1. \end{cases} \tag{3.189}$$

This corresponds to compressing the unit square in the p direction and stretching it in the q direction, then cutting it along the p direction and finally stacking one piece on top of the other (similarly to the way a baker kneads dough). Note that map (3.189) is area preserving. The baker's map is a paradigmatic model of classical chaos. Indeed, it exhibits sensitive dependence on initial conditions, which is the distinctive feature of classical chaos: any small error in determining the initial conditions is amplified exponentially in time. In other words, two nearby trajectories separate exponentially, with a rate given by the maximum Lyapunov exponent λ, defined as

$$\lambda = \lim_{|t| \to \infty} \frac{1}{t} \ln\left(\frac{\delta(t)}{\delta(0)}\right), \tag{3.190}$$

where the discrete time t measures the number of map iterations and $\delta(t) = \sqrt{(\delta q(t))^2 + (\delta p(t))^2}$. To compute $\delta q(t)$ and $\delta p(t)$, we differentiate map (3.189), obtaining

$$\begin{bmatrix} \delta\bar{q} \\ \delta\bar{p} \end{bmatrix} = M \begin{bmatrix} \delta q \\ \delta p \end{bmatrix} = \begin{bmatrix} 2 & 0 \\ 0 & \frac{1}{2} \end{bmatrix} \begin{bmatrix} \delta q \\ \delta p \end{bmatrix}. \tag{3.191}$$

The iteration of map (3.191) gives $\delta q(t)$ and $\delta p(t)$ as a function of $\delta q(0)$ and $\delta p(0)$. The matrix M, called the stability matrix, has eigenvalues $\mu_1 = 2$ and $\mu_2 = \frac{1}{2}$, which do not depend on the coordinates q and p. In this case, the maximum Lyapunov exponent is simply given by $\lambda = \ln \mu_1 = \ln 2$. The

consists of the iteration of a unitary operator. Each step of this iteration maps the state of the quantum computer onto a new state. If this discrete-time dynamics is repeated for a suitable number of iterations, the probability of detecting the marked item becomes of order unity. The dynamics associated with Grover's algorithm has an interesting phase-space representation, discussed in Miquel *et al.* (2002b).

dynamics is uniformly unstable, inducing contraction in the p direction and stretching in the q direction.

The baker's map can be quantized following Balazs and Voros (1989) and Saraceno (1990). We introduce the position (q) and momentum (p) operators, and denote the eigenstates of these operators by $|q_j\rangle$ and $|p_k\rangle$, respectively. The corresponding eigenvalues are given by $q_j = j/N$ and $p_k = k/N$, with $j, k = 0, \ldots, N - 1$, N being the dimension of the Hilbert space. Note that, to fit N levels onto the unit square, we must set $2\pi\hbar = 1/N$. We take $N = 2^n$, where n is the number of qubits used to simulate the quantum baker's map on a quantum computer. The transformation between the position basis $\{|q_0\rangle, \ldots, |q_{N-1}\rangle\}$ and the momentum basis $\{|p_0\rangle, \ldots, |p_{N-1}\rangle\}$ is performed by means of a discrete Fourier transform F_n, defined by the matrix elements

$$\langle q_k|F_n|q_j\rangle \equiv \frac{1}{\sqrt{2^n}} \exp\left(\frac{2\pi i k j}{2^n}\right). \tag{3.192}$$

It can be shown that the quantized baker's map may be defined by the transformation (see Balazs and Voros, 1989)

$$|\psi\rangle \rightarrow |\bar{\psi}\rangle = T|\psi\rangle, \tag{3.193}$$

where $|\bar{\psi}\rangle$ denotes the wave vector of the system after application of one map step to the state $|\psi\rangle$. The unitary transformation T defining the quantum baker's map is given by

$$T = F_n^{-1} \begin{bmatrix} F_{n-1} & 0 \\ 0 & F_{n-1} \end{bmatrix}, \tag{3.194}$$

where the matrix elements are to be understood relative to the position basis $\{|q_j\rangle\}$ and F_{n-1} is the discrete Fourier transform defined by Eq. (3.192).

The unitary transformation T can be implemented on a quantum computer with n qubits. We have

$$T = F_n^{-1} (I \otimes F_{n-1}), \tag{3.195}$$

where F_{n-1} acts on the $n - 1$ least significant qubits and I is the identity operator acting on the most significant qubit. Since the Fourier transform F_n can be implemented efficiently on a quantum computer with $O(n^2)$ elementary gates, the quantum baker's map can also be simulated efficiently on a quantum computer with $O(n^2)$ elementary gates per map iteration.

Note that the best known classical algorithm to simulate the Fourier transform, the fast Fourier transform, requires $O(n2^n)$ elementary operations. Therefore, the dynamics of the baker's map can be simulated on a quantum computer with an exponential gain with respect to any known classical computation.

3.15.3 * The quantum sawtooth map

The sawtooth map is a prototype model in the studies of classical and quantum-dynamical systems and exhibits a rich variety of interesting physical phenomena, from complete chaos to complete integrability, normal and anomalous diffusion, dynamical localization and cantori localization. Furthermore, the sawtooth map gives a good approximation to the motion of a particle bouncing inside a stadium billiard (which is a well-known model of classical and quantum chaos). In this subsection, we show that this map can be efficiently simulated on a quantum computer.

The sawtooth map belongs to the class of periodically driven dynamical systems, governed by the Hamiltonian

$$H(\theta, I; \tau) = \frac{I^2}{2} + V(\theta) \sum_{j=-\infty}^{+\infty} \delta(\tau - jT), \qquad (3.196)$$

where (I, θ) are conjugate action-angle variables ($0 \le \theta < 2\pi$). This Hamiltonian is the sum of two terms, $H(\theta, I; \tau) = H_0(I) + U(\theta; t)$, where $H_0(I) = I^2/2$ is just the kinetic energy of a free rotator (a particle moving on a circle parametrized by the coordinate θ), while

$$U(\theta; t) = V(\theta) \sum_j \delta(\tau - jT) \qquad (3.197)$$

represents a force acting on the particle that is switched on and off instantaneously at time intervals T. Therefore, we say that the dynamics described by Hamiltonian (3.196) is *kicked*. The corresponding Hamiltonian equations of motion are

$$\begin{cases} \dot{I} = -\dfrac{\partial H}{\partial \theta} = -\dfrac{dV(\theta)}{d\theta} \displaystyle\sum_{j=-\infty}^{+\infty} \delta(\tau - jT), \\[2mm] \dot{\theta} = \dfrac{\partial H}{\partial I} = I. \end{cases} \qquad (3.198)$$

These equations can be easily integrated and one finds that the evolution from time lT^- (prior to the l-th kick) to time $(l+1)T^-$ (prior to the

$(l+1)$-th kick) is described by the map

$$\begin{cases} \bar{I} = I + F(\theta), \\ \bar{\theta} = \theta + T\bar{I}, \end{cases} \tag{3.199}$$

where $F(\theta) = -dV(\theta)/d\theta$ is the force acting on the particle.

In the following, we shall consider the special case $V(\theta) = -k(\theta - \pi)^2/2$. This map is called the *sawtooth map*, since the force $F(\theta) = -dV(\theta)/d\theta = k(\theta - \pi)$ has a sawtooth shape, with a discontinuity at $\theta = 0$. By rescaling $I \to J = TI$, the classical dynamics is seen to depend only on the parameter $K = kT$. Indeed, in terms of the variables (J, θ) map (3.199) becomes

$$\begin{cases} \bar{J} = J + K(\theta - \pi), \\ \bar{\theta} = \theta + \bar{J}. \end{cases} \tag{3.200}$$

A stability analysis of the sawtooth map can be performed analogously to the case of the baker's map, discussed in the previous subsection. After differentiating map (3.200), we have

$$\begin{bmatrix} \delta\bar{J} \\ \delta\bar{\theta} \end{bmatrix} = M \begin{bmatrix} \delta J \\ \delta\theta \end{bmatrix} = \begin{bmatrix} 1 & K \\ 1 & 1+K \end{bmatrix} \begin{bmatrix} \delta J \\ \delta\theta \end{bmatrix}. \tag{3.201}$$

The stability matrix M has eigenvalues $\mu_\pm = \frac{1}{2}(2 + K \pm \sqrt{K^2 + 4K})$, which are complex conjugate for $-4 \leq K \leq 0$ and real for $K < -4$ and $K > 0$. Thus, the classical motion is stable for $-4 \leq K \leq 0$ and completely chaotic for $K < -4$ and $K > 0$.[6]

The sawtooth map can be studied on the cylinder ($J \in (-\infty, +\infty)$), which can also be closed to form a torus of length $2\pi L$ (L is an integer, to assure that no discontinuities are introduced in the second equation of (3.200) when J is taken modulo $2\pi L$). For any $K > 0$, one has normal diffusion in the action (momentum) variable. Although the sawtooth map is a deterministic system, the motion of a trajectory along the momentum direction is in practice indistinguishable from a random walk. Thus, the evolution of a distribution function $f(J, t)$ is governed by a Fokker–Planck equation:

$$\frac{\partial f}{\partial t} = \frac{\partial}{\partial J}\left(\frac{1}{2}D\frac{\partial f}{\partial J}\right). \tag{3.202}$$

---

[6]The maximum Lyapunov exponent is $\lambda = \ln \mu_+$ for $K > 0$, $\lambda = \ln|\mu_-|$ for $K < -4$ and $\lambda = 0$ in the stable region $-4 \leq K \leq 0$.

Here $t \equiv \tau/T$ is the discrete time measured in units of map iterations and the diffusion coefficient D is defined by

$$D = \lim_{t \to \infty} \frac{\langle (\Delta J(t))^2 \rangle}{t}. \tag{3.203}$$

where $\Delta J \equiv J - \langle J \rangle$ and $\langle \ldots \rangle$ denotes the average over an ensemble of trajectories. If at time $t = 0$ we take a phase space distribution with initial momentum J_0 and random phases $0 \le \theta < 2\pi$, then the solution of the Fokker–Planck equation (3.202) is given by

$$f(J,t) = \frac{1}{\sqrt{2\pi D t}} \exp\left(-\frac{(J - J_0)^2}{2Dt}\right). \tag{3.204}$$

The width of this Gaussian distribution grows in time, according to

$$\langle (\Delta J(t))^2 \rangle \approx D(K)\, t. \tag{3.205}$$

For $K > 1$, the diffusion coefficient is well approximated by the random phase approximation, in which we assume that there are no correlations between the angles (phases) θ at different times. Hence, we have

$$D(K) \approx \frac{1}{2\pi} \int_0^{2\pi} d\theta\, (\Delta J_1)^2 = \frac{1}{2\pi} \int_0^{2\pi} d\theta\, K^2 (\theta - \pi)^2 = \frac{\pi^2}{3} K^2, \tag{3.206}$$

where $\Delta J_1 = \bar{J} - J$ is the change in action after a single map step. For $0 < K < 1$ diffusion is slowed, owing to the sticking of trajectories close to broken tori (known as cantori), and we have $D(K) \approx 3.3\, K^{5/2}$ (this regime is discussed in Dana *et al.*, 1989). For $-4 < K < 0$ the motion is stable, the phase space has a complex structure of elliptic islands down to smaller and smaller scales and one can observe anomalous diffusion, that is, $\langle (\Delta J)^2 \rangle \propto t^\alpha$, with $\alpha \ne 1$ (for instance, $\alpha = 0.57$ when $K = -0.1$). The cases $K = -1, -2, -3$ are integrable.

The quantum version of the sawtooth map is obtained by means of the usual quantization rules, $\theta \to \theta$ and $I \to I = -i\partial/\partial\theta$ (we set $\hbar = 1$). The quantum evolution in one map iteration is described by a unitary operator U acting on the wave function ψ:

$$\bar{\psi} = U\psi = \exp\left(-i \int_{lT^-}^{(l+1)T^-} d\tau\, H(\theta, I; \tau)\right)\psi, \tag{3.207}$$

where H is Hamiltonian (3.196). Since the potential $V(\theta)$ is switched on only at discrete times lT, it is straightforward to obtain

$$\bar{\psi} = e^{-iTI^2/2} e^{-iV(\theta)} \psi = e^{-iTI^2/2} e^{ik(\theta-\pi)^2/2} \psi. \qquad (3.208)$$

The effective Planck constant is given by $\hbar_{\text{eff}} = \hbar/k$ and the classical limit corresponds to $k \to \infty$ and $T \to 0$ while keeping $K = kT$ constant.

In the following, we describe an exponentially efficient quantum algorithm for simulation of the map (3.208). It is based on the forward/backward quantum Fourier transform between momentum and angle bases. Such an approach is convenient since the operator U, introduced in Eq. (3.207), is the product of two operators, $U_k = e^{ik(\theta-\pi)^2/2}$ and $U_T = e^{-iTI^2/2}$, diagonal in the θ and I representations, respectively. This quantum algorithm requires the following steps for one map iteration:

1. We apply U_k to the wave function $\psi(\theta)$. In order to decompose the operator U_k into one- and two-qubit gates, we first of all write θ in binary notation:

$$\theta = 2\pi \sum_{j=1}^{n} \alpha_j 2^{-j}, \qquad (3.209)$$

with $\alpha_i \in \{0,1\}$. Here n is the number of qubits, so that the total number of levels in the quantum sawtooth map is $N = 2^n$. From this expansion, we obtain

$$(\theta - \pi)^2 = 4\pi^2 \sum_{j_1,j_2=1}^{n} \left(\frac{\alpha_{j_1}}{2^{j_1}} - \frac{1}{2n}\right)\left(\frac{\alpha_{j_2}}{2^{j_2}} - \frac{1}{2n}\right). \qquad (3.210)$$

This term can be put into the unitary operator U_k, giving the decomposition

$$e^{ik(\theta-\pi)^2/2} = \prod_{j_1,j_2=1}^{n} \exp\left[i2\pi^2 k \left(\frac{\alpha_{j_1}}{2^{j_1}} - \frac{1}{2n}\right)\left(\frac{\alpha_{j_2}}{2^{j_2}} - \frac{1}{2n}\right)\right], \qquad (3.211)$$

which is the product of n^2 two-qubit gates (controlled phase-shift gates), each acting non-trivially only on the qubits j_1 and j_2. In the computational basis $\{|\alpha_{j_1}\alpha_{j_2}\rangle = |00\rangle, |01\rangle, |10\rangle, |11\rangle\}$ each two-qubit gate can be written as $\exp(i2\pi^2 k D_{j_1,j_2})$, where D_{j_1,j_2} is a diagonal matrix:

$$D_{j_1,j_2} = \begin{bmatrix} \frac{1}{4n^2} & 0 & 0 & 0 \\ 0 & -\frac{1}{2n}\left(\frac{1}{2^{j_2}} - \frac{1}{2n}\right) & 0 & 0 \\ 0 & 0 & -\frac{1}{2n}\left(\frac{1}{2^{j_1}} - \frac{1}{2n}\right) & 0 \\ 0 & 0 & 0 & \left(\frac{1}{2^{j_1}} - \frac{1}{2n}\right)\left(\frac{1}{2^{j_2}} - \frac{1}{2n}\right) \end{bmatrix}.$$

(3.212)

2. The change from the θ to the I representation is obtained by means of the quantum Fourier transform, which, as we know, requires n Hadamard gates and $\frac{1}{2}n(n-1)$ controlled phase-shift gates.

3. In the I representation, the operator U_T has essentially the same form as the operator U_k in the θ representation and can therefore be decomposed into n^2 controlled phase-shift gates, similarly to Eq. (3.211).

4. We return to the initial θ representation by application of the inverse quantum Fourier transform.

Thus, overall, this quantum algorithm requires $3n^2 + n$ gates per map iteration ($3n^2 - n$ controlled phase-shifts and $2n$ Hadamard gates). This number is to be compared with the $O(n2^n)$ operations required by a classical computer to simulate one map iteration by means of a fast Fourier transform. Thus, the quantum simulation of the quantum sawtooth map dynamics is exponentially faster than any known classical algorithm. Note that the resources required to the quantum computer to simulate the evolution of the sawtooth map are only logarithmic in the system size N. Of course, there remains the problem of extracting useful information from the quantum-computer wave function. This will be discussed in the next subsection.

3.15.4   * *Quantum computation of dynamical localization*

Dynamical localization is one of the most interesting phenomena that characterize the quantum behaviour of classically chaotic systems: quantum interference effects suppress chaotic diffusion in momentum, leading to exponentially localized wave functions. This phenomenon was first found and studied in the quantum kicked-rotator model and has profound analogies with Anderson localization of electronic transport in disordered materials. Dynamical localization has been observed experimentally in the microwave ionization of Rydberg atoms and in experiments with cold atoms.

Dynamical localization can be studied in the sawtooth map model. In this case, map (3.208) is studied on the cylinder ($I \in (-\infty, +\infty)$), which is cut-off at a finite number N of levels due to the finite quantum (or classical) computer memory. Similarly to other models of quantum chaos, quantum interference in the sawtooth map leads to suppression of classical chaotic diffusion after a *break time*

$$t^\star \approx D \approx (\pi^2/3)k^2 \,, \tag{3.213}$$

where D is the classical diffusion coefficient, measured in number of levels ($\langle (\Delta m)^2 \rangle \approx Dt$, where the index m singles out the eigenstates of I, that is, $I|m\rangle = m|m\rangle$). After the break time $t^\star$, the probability distribution over the momentum eigenbasis *decays exponentially*:

$$W_m \equiv \left| \langle m|\psi\rangle \right|^2 \approx \frac{1}{\ell} \exp\left[-\frac{2|m - m_0|}{\ell} \right], \tag{3.214}$$

with m_0 the initial value of the momentum. Therefore, for $t > t^\star$ only $\sqrt{\langle (\Delta m)^2 \rangle} \sim \ell$ levels are populated and the localization length ℓ for the average probability distribution is approximately equal to the classical diffusion coefficient:

$$\ell \approx D. \tag{3.215}$$

Thus, the quantum localization can be detected if ℓ is smaller than the system size N.

In Fig. 3.27, we show that exponential localization can already be clearly seen with $n = 6$ qubits.

It is useful to stress again that in a quantum computer the memory capabilities grow exponentially with the number of qubits (the number of levels N is equal to 2^n). Therefore, already with less than 40 qubits, one could make simulations inaccessible to today's supercomputers. After the break time $t^\star$, the decay of the probability distribution over the momentum eigenbasis is indeed exponential. Fig. 3.27 shows that the exponentially localized distribution, appearing at $t \approx t^\star$, is *frozen* in time, apart from quantum fluctuations, which we partially smooth out by averaging over a few map steps. The freezing of the localized distribution can be seen from comparison of the probability distributions taken immediately after $t^\star$ (the full curve in Fig. 3.27) and at a much larger time $t = 300 \approx 25t^\star$ (the dashed curve in the same figure). Here the localization length is $\ell \approx 12$ and classical diffusion is suppressed after a break time $t^\star \approx \ell \approx D$ (the diffusion coefficient is $D \approx (\pi^2/3)k^2 \approx 9.9$), in agreement with estimates

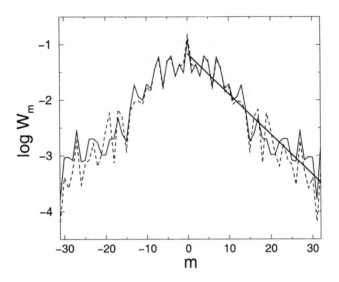

Fig. 3.27   The probability distribution over the momentum basis for the sawtooth map
with $n = 6$ qubits, $k = \sqrt{3}$, $K = \sqrt{2}$ and initial momentum $m_0 = 0$; the time average is
taken in the intervals $10 \leq t \leq 20$ (full curve) and $290 \leq t \leq 300$ (dashed curve). The
straight line fit, $W_m \propto \exp(-2|m|/\ell)$, gives a localization length $\ell \approx 12$. Note that the
logarithm is base ten. Figure taken from Benenti *et al.* (2003).

(3.213–3.215). This quantum computation up to times of the order of ℓ
requires a number $N_g \approx 3n^2\ell \sim 10^3$ of one- or two-qubit quantum gates.

We now discuss how to extract information (the value of the localization
length) from a quantum computer simulating the sawtooth-map dynamics.
The localization length can be measured by running the algorithm repeat-
edly up to time $t > t^*$. Each run is followed by a standard projective
measurement on the computational (momentum) basis. Since the wave
function at time t can be written as

$$|\psi(t)\rangle = \sum_m \hat{\psi}(m,t)\,|m\rangle, \qquad (3.216)$$

with $|m\rangle$ momentum eigenstates, such a measurement gives outcome $\bar{m}$
with probability

$$W_{\bar{m}} = \left|\langle \bar{m}|\psi(t)\rangle\right|^2 = \left|\hat{\psi}(\bar{m},t)\right|^2. \qquad (3.217)$$

A first series of measurements would allow us to give a rough estimate of the
variance $\langle(\Delta m)^2\rangle$ of the distribution W_m. In turn, $\sqrt{\langle(\Delta m)^2\rangle}$ gives a first
estimate of the localization length ℓ. After this, we can store the results

of the measurements in histogram bins of width $\delta m \propto \ell \approx \sqrt{\langle (\Delta m)^2 \rangle}$. Finally, the localization length is extracted from a fit of the exponential decay of this coarse-grained distribution over the momentum basis. Elementary statistical theory tells us that, in this way, the localization length can be obtained with accuracy ν after the order of $1/\nu^2$ computer runs. It is interesting to note that it is sufficient to perform a coarse-grained measurement to generate a coarse-grained distribution. This means that it will be sufficient to measure the most significant qubits and ignore those that would give a measurement accuracy below the coarse graining δm. Thus, the number of runs and measurements is independent of ℓ.

We now come to the crucial point of estimating the gain of quantum computation of the localization length with respect to classical computation. First of all, we recall that it is necessary to make about $t^\star \sim \ell$ map iterations to obtain the localized distribution, see Eqs. (3.213) and (3.215). This is true, both for the present quantum algorithm and for classical computation. It is reasonable to use a basis size $N \propto \ell$ to detect localization (say, N equal to a few times the localization length). In such a situation ($N \sim \ell$), a classical computer requires $O(\ell^2 \log \ell)$ operations to extract the localization length, while a quantum computer would require $O(\ell(\log \ell)^2)$ elementary gates. In this sense, for $\ell \sim N = 2^n$ the quantum computer provides a *quadratic speed up*, since both classical and quantum computers perform $O(N)$ map iterations. However, for a fixed number of iterations t, the quantum computation provides an *exponential gain* since in this case one should compare $O(t(\log N)^2)$ gates (quantum computation) with $O(tN \log N)$ gates (classical computation).

The simulation of quantum dynamics up to a given time t is useful, for instance, to measure dynamical correlation functions of the form

$$C(t) \equiv \langle \psi | A^\dagger(t) B(0) | \psi \rangle = \langle \psi | (U^\dagger)^t A^\dagger(0) U^t B(0) | \psi \rangle, \qquad (3.218)$$

where U is the time-evolution operator (3.207) for the sawtooth map. Similarly, we can efficiently compute the fidelity of quantum motion, which is a quantity of central interest in the study of the stability of quantum motion under perturbations. The fidelity $f(t)$ (also called the Loschmidt echo), measures the accuracy with which a quantum state can be recovered by inverting, at time t, the dynamics with a perturbed Hamiltonian. It is defined as

$$f(t) = \langle \psi | (U_\epsilon^\dagger)^t U^t | \psi \rangle. \qquad (3.219)$$

Here the wave vector $|\psi\rangle$ evolves forward in time with the Hamiltonian H of Eq. (3.196) up to time t and then evolves backward in time with a perturbed Hamiltonian H_ϵ (U_ϵ is the corresponding time-evolution operator). For instance, we may perturb the parameter k in the sawtooth map as follows: $k \rightarrow k' = k + \epsilon$. If the evolution operators U and U_ϵ can be simulated efficiently on a quantum computer, as is the case in most physically interesting situations, then the fidelity of quantum motion can be evaluated with exponential speed up with respect to known classical computations. The same conclusion is valid for the correlation functions (3.218).

3.16 First experimental implementations

The great challenge of quantum computation is to experimentally realize a quantum computer. Many requirements must be fulfilled in order to achieve this imposing objective. We require a collection of two-level quantum system that can be prepared, manipulated and measured at will. That is, our purpose is to be able to control and measure the state of a many-qubit quantum system. A useful quantum computer must be *scalable* since we need a rather large number of qubits to perform non-trivial computations. In other words, we need the quantum analogue of the integrated circuits of a classical computer. As will become clear in the next subsection, qubits must interact in a controlled way if we wish to be able to implement a universal set of quantum gates. Furthermore, we must be able to control the evolution of a large number of qubits for the time necessary to perform many quantum gates.

As we shall discuss in detail in Chap. 6, decoherence may be considered as the ultimate obstacle to the practical realization of a quantum computer. Here the term decoherence (or loss of quantum coherence) denotes the corruption of the quantum information stored in the quantum computer, due to the unavoidable coupling of the quantum computer to the surrounding environment. This coupling affects the performance of a quantum computer, introducing errors in the computation. Another source of errors that must be taken into account is the presence of imperfections in the quantum computer hardware.

As we shall discuss in Chap. 7, quantum error correcting codes exist. However, a necessary requirement for a successful correction procedure is that one may implement many quantum gates inside the decoherence time scale. Here "many" means $10^3 - 10^4$, the exact value depending on the kind of error. It should be appreciated that it appears very hard to fulfil all the above requirements in interacting many-qubit quantum systems. It is also important to note that the quantum computer must be well isolated from the environment to protect it from errors, but at the same time easily accessible, as we wish to manipulate its state (preparation, controlled evolution and read out). It is a non-trivial task to combine these two conflicting requirements since the external control operations generally introduce undesired coupling to the environment, for instance due to noise in quantum gates.

3.16.1 *Elementary gates with spin qubits*

To understand the basic physical principles underlying any experimental implementation of one- and two-qubit quantum gates, it is instructive to consider the following simple example. Let us assume that we have an isolated spin-$\frac{1}{2}$ particle (our qubit) in the presence of a static plus a time-dependent magnetic field. The particle could be an electron or a nuclear spin. Such a system is described by the Hamiltonian

$$H = -\mu\Big\{H_0\sigma_z + H_1\big[\cos(\omega t)\sigma_x + \sin(\omega t)\sigma_y\big]\Big\}, \tag{3.220}$$

where H_0 and H_1 are the strengths of the static and the oscillating magnetic fields, respectively. Note that the static field is directed along z, while the oscillating field lies on the (x, y) plate and rotates uniformly about the z axis.

The evolution of the state $|\psi(t)\rangle$ of the spin-$\frac{1}{2}$ particle can be computed analytically (see exercise 3.22 below). The resonance condition $\omega = \omega_0 \equiv -2\mu H_0/\hbar$ is particularly interesting. This condition is satisfied when the angular frequency of the oscillating field (multiplied by $\hbar$) is equal to the energy difference between the two spin states (in the presence of the static field only), namely, when $\hbar\omega = -2\mu H_0$. In this case, the solution writes

$$|\psi(t)\rangle = U|\psi(0)\rangle = e^{-i\omega\sigma_z t/2}\,e^{-i\Omega\sigma_x t/2}\,|\psi(0)\rangle, \tag{3.221}$$

where $\Omega = -2\mu H_1/\hbar$ is the Rabi frequency. Following exercise 2.12, we can write the unitary evolution operator U of Eq. (3.221) in the computational basis. We obtain

$$U = \begin{bmatrix} e^{-i\omega t/2} & 0 \\ 0 & e^{i\omega t/2} \end{bmatrix} \begin{bmatrix} \cos(\Omega t/2) & -i\sin(\Omega t/2) \\ i\sin(\Omega t/2) & \cos(\Omega t/2) \end{bmatrix}. \tag{3.222}$$

Exercise 3.22　Solve the Schrödinger equation

$$i\hbar\frac{d}{dt}|\psi(t)\rangle = -\mu\{H_0\sigma_z + H_1[\cos(\omega t)\sigma_x + \sin(\omega t)\sigma_y]\}|\psi(t)\rangle. \quad (3.223)$$

In particular, discuss the resonance condition $\omega = -2\mu H_0/\hbar$.

The application, for a given period of time τ, of an oscillating magnetic field satisfying the resonance condition $\omega = \omega_0$, is called a *Rabi pulse*. It is important to underline that Rabi pulses of appropriate duration implement single-qubit quantum gates. For instance, let us consider a pulse of period τ such that $\Omega\tau = \pi$. It is easy to see from (3.222) that this pulse reproduces a NOT gate, up to phase factors. Indeed, if the system is initially in the state $|0\rangle$, it ends up in the state

$$ie^{i\omega\tau/2}|1\rangle. \quad (3.224)$$

In contrast, if the system is initially in the state $|1\rangle$, it ends up in the state

$$-ie^{-i\omega\tau/2}|0\rangle. \quad (3.225)$$

In order to perform exactly a NOT gate, we must eliminate the phase factors, by setting $\omega\tau = (4n + 3)\pi$. Similarly, we can produce arbitrary single-qubit unitary transformations.

So far, we have discussed single-qubit gates. However, the realization of two-qubit controlled gates is a necessary requirement for the implementation of universal quantum computation. In is important to point out that, for this purpose, we need *interacting* qubits. The following simple example will help clarify this concept. Let us consider a model of two coupled spin-$\frac{1}{2}$ particles, described by the Hamiltonian

$$H(t) = H_s + H_p(t), \quad (3.226)$$

where

$$H_s = -\left(\mu_1\sigma_z^{(1)} + \mu_2\sigma_z^{(2)}\right)H_0 + J\sigma_z^{(1)}\sigma_z^{(2)} \quad (3.227)$$

and $H_p(t)$ is a time-dependent Hamiltonian describing a pulse suitable to realize a controlled gate. The first two terms in H_s describe the effect of the static magnetic field H_0 on the two particles, while $J\sigma_z^{(1)}\sigma_z^{(2)}$ represents an Ising interaction between the qubits. The Hamiltonian H_s describes a conservative system, whose eigenstates are the states of the computational basis, namely, $|00\rangle$, $|01\rangle$, $|10\rangle$ and $|11\rangle$, and the corresponding eigenvalues are given by

$$E_{00} = -(\mu_1 + \mu_2)H_0 + J\,, \qquad E_{01} = -(\mu_1 - \mu_2)H_0 - J\,,$$
$$E_{10} = (\mu_1 - \mu_2)H_0 - J\,, \qquad E_{11} = (\mu_1 + \mu_2)H_0 + J\,. \qquad (3.228)$$

Let us assume that we wish to implement a CNOT gate. This is not possible in the non-interacting case $J = 0$. Indeed, if we apply a resonant in this case pulse, that is, an oscillating magnetic field with frequency $\omega = -2\mu_2 H_0/\hbar$, we induce the transition $|0\rangle \leftrightarrow |1\rangle$ on the second qubit, *independently* of the state of the first qubit. The resonance condition is satisfied whatever the state of the first qubit is. In contrast, in the interacting case $J \neq 0$, we can implement a CNOT gate if the oscillating magnetic field has frequency $\omega(J) = -2(\mu_2 H_0 + J)/\hbar$. Indeed, the resonance condition is satisfied for the transition $|10\rangle \leftrightarrow |11\rangle$ but not for the transition $|00\rangle \leftrightarrow |01\rangle$. In the first case, the energy difference between the two levels involved is $-2(\mu_2 H_0 + J)$ while in the latter it is $-2(\mu_2 H_0 - J)$.

3.16.2 *Overview of the first implementations*

This subsection gives a short overview of the first experimental implementations of quantum logic gates and few-qubit quantum processors. A much more detailed discussion of this topic is postponed until Chap. 8. Given the generality of the requirements to build a quantum computer, many physical systems might be good candidates. Let us briefly discuss a few of them. Note that this is by no means an exhaustive list. Moreover, given the state-of-the-art technology, some of the proposals discussed below are less realistic then others. However, we present them to give a flavour of the rich variety of physical systems that are under active investigation.

Liquid-state NMR. The quantum hardware consists of a liquid containing a large number (of order 10^{18}) of molecules of a given type, placed in a strong static magnetic field. A qubit is the spin of a nucleus in a molecule and quantum gates are implemented by means of resonant oscillating magnetic fields (Rabi pulses), that is, nuclear magnetic resonance (NMR) techniques are used. Quantum information exchange between nuclei inside a molecule is based on spin-spin interactions (chemical bonds) between neighbouring atoms. The molecules are prepared in thermal equilibrium at room temperature. It is important to stress that in liquid-state NMR the spin state of a single nucleus is neither prepared nor measured. On the contrary, we measure the average spin state of the $\sim 10^{18}$ molecules contained in the solution. The fast chaotic rotational motion of the molecules in the liquid state allows us to neglect the intermolecular interactions, which average to zero over a time scale much shorter than the time required to implement a quantum gate. Therefore, we can consider the $\sim 10^{18}$ molecules in the solution as quantum processors evolving independently from each other.

With NMR experiments, it has been possible to experimentally demonstrate several quantum algorithms, including Grover's algorithm, the quantum Fourier transform and the quantum baker's map. These algorithms have been implemented on three-qubit molecules. Moreover, the simplest instance of Shor's algorithm, namely, the factorization of 15, has been implemented using a seven-qubit molecule and a sequence of about 300 Rabi pulses. Unfortunately, liquid-state NMR quantum computing is not scalable since the measured signal drops exponentially with the number of qubits in a molecule.

Optical systems. A qubit can be realized with a single photon in two optical modes, such as horizontal or vertical polarization states. Single-qubit gates can be implemented using linear-optics devices such as beam splitters and phase shifters. An interaction between photons is possible, but technically difficult, as it must be mediated by atoms in a non-linear Kerr medium. However, Knill, Laflamme and Milburn have recently shown that linear optics is sufficient to perform quantum computation with photons efficiently, provided that measurements (via photo-detectors) can be performed at any time during a quantum computation and that the measurement result can be used to control other optical elements.

Cavity quantum electrodynamics. Using cavity quantum electrodynamics (QED) techniques, it has been possible to realize experiments in which a single atom interacts with a single mode or a few modes of the electromagnetic field inside a cavity. The two states of a qubit can be represented by the polarization states of a single photon or by two excited states of an atom. In the first case, the interaction between qubits is mediated by atoms, in the latter case by photons. Cavity QED techniques have allowed the implementation of one- and two-qubit gates; however, it seems very difficult to perform a large number of operations or to address the scalability problem with these techniques. Nevertheless, it should be stressed that cavity QED experiments have been particularly successful in demonstrating basic features of quantum mechanics, such as Rabi oscillations and entanglement, or in exploring the effect of decoherence and the transition from the quantum world to classical physics.

Ion-traps. The quantum hardware is as follows: a string of ions is confined by a combination of static and oscillating electric fields in a linear trap (known as a Paul trap). A qubit is a single ion and two long-lived states of the ion correspond to the two states of the qubit. The linear array of ions held in the trap is the quantum register. Single-qubit gates are obtained by addressing individual ions with laser Rabi pulses. The interactions between qubits, which are necessary to implement controlled two-qubit operations, are mediated by the collective vibrational motion of the trapped string of ions. The ion-trap technique has allowed the implementation of basic one- and two-qubit gates, the entanglement of four ions and, very recently, the demonstration of the Deutsch–Jozsa algorithm. This latter result shows the capability of the ion-trap technique to realize several quantum gates within the relevant decoherence time scales. It seems probable that in the next few years it will become possible to apply tens of quantum gates to a few ions without loosing quantum coherence.

To build a scalable quantum computer, Cirac and Zoller envisaged a two-dimensional array of independent ion traps and an independent ion (head) that moves about this plane, capable of approaching any particular ion. A suitable laser pulse could swap the state of the ion into the head and this would allow us to entangle well-separated ions, the quantum communication between them being assured by the moving head. It seems that there are no fundamental physical obstacles against this proposal, but a significant technological challenge remains.

Solid state proposals. Several proposals have been put forward to build a solid-state quantum computer. This is not surprising, since solid-state physics has developed over the years a sophisticated technology, creating artificial structures and devices on nanoscale. Solid-state physics is at the basis of the development of classical computer technology and therefore the scalability problem would find a natural solution in a solid state quantum computer. Indeed, such a quantum computer could benefit from the fabrication techniques of microelectronics.

In the remaining part of this section, we shall briefly discuss three solid-state proposals.

Quantum dots. Quantum dots are structures fabricated from semiconductor materials, in which electrostatic potentials confine electrons. The dot size is typically between 10 nanometres and 1 micron. The qubit is realized as the spin of an electron on a single-electron quantum dot and two-qubit quantum operations are operated by a purely electrical gating of the electrostatic tunnelling barrier between neighbouring quantum dots. Lowering (raising) this barrier correspond to switching on (off) the interaction between two qubits. Scalability is in principle possible, since it is possible to produce arrays of quantum dots with present technology. However, the actual implementation of quantum gates and single spin measurements in such arrays constitutes a difficult experimental challenge. Furthermore, there are a great variety of possible decoherence processes in complex solid-state devices and our knowledge of them is still very limited.

Spins in semiconductors. A proposal by Kane combines solid state NMR techniques with silicon microchip technology. The idea is to place single phosphorus atoms in a silicon matrix. The qubit is the nuclear spin-$\frac{1}{2}$ of a single phosphorus atom. The interaction between qubits is mediated by the hyperfine interaction between the qubits and the surrounding electrons. Electric gates control the individual electronic states and the interactions between qubits, while magnetic fields (Rabi pulses) implement quantum logic operations. This proposal is well beyond state-of-the-art technology, since it requires nanofabrication on the atomic scale. To grasp the difficulties of this proposal, note that the separation between electric gates should be of the order of one micron and phosphorus atoms should be placed in an ordered array on the same length scale. In any case, we should also remember that silicon technology is a very rapidly developing field.

Superconducting circuits. Recently there has been very remarkable experimental progress using superconducting microelectronic circuits to construct artificial two-level systems. In superconductors, pairs of electrons are bound together to form objects of charge twice the electron charge, called Cooper pairs. Electrostatic potentials can confine the Cooper pairs in a "box" of micron size. Two charge states of a box, differing by one Cooper pair, can be used to represent the two states of a qubit. Josephson-junction circuits are used to control the state of the qubit. In a Josephson junction two Cooper-pair boxes are separated by a thin insulator and a Cooper pair can move from one box to the other by a quantum tunnelling effect. It is possible to induce Rabi oscillations between these two states, thus implementing single-qubit logic gates. In another approach a magnetic flux is applied to a superconducting ring and the two states of a qubit correspond to clockwise and counterclockwise circulating currents. With both approaches, it has been possible to observe Rabi oscillations and to control the state of the qubit by means of microwave Rabi pulses. Moreover, it has been possible to couple superconducting qubits and to observe the effect of this coupling on the Rabi oscillations for a qubit. Recently, conditional gate operation has been demonstrated using a pair of coupled superconducting qubits.

3.17   A guide to the bibliography

The quantum Turing machine is discussed, *e.g.*, in Galindo and Martin-Delgado (2002). A pioneering study of the circuit model of quantum computation was given by Deutsch (1989). A clear discussion of the role of entanglement in quantum computational speed-up is Jozsa and Linden (2002), see also Biham *et al.* (2003).

The universality of two-qubit quantum gates is discussed in Reck *et al.* (1994), DiVincenzo (1995), Barenco (1995), Deutsch *et al.* (1995) and Lloyd (1995). Many useful circuit constructions can be found in Barenco *et al.* (1995) and Song and Klappenecker (2002). The decomposition of unitary matrices into matrices of smaller size is discussed by Tucci (1999).

A practical method of constructing quantum logic circuits is discussed in Lee *et al.* (1999). Useful quantum circuits implementing various arithmetic operations can be found in Vedral *et al.* (1996), Beckman *et al.* (1996), Miquel *et al.* (1996), Gossett (1998) and Draper (2000).

Deutsch's algorithm was invented by Deutsch (1985) and extended to the n-qubit case by Deutsch and Jozsa (1992), see also Cleve *et al.* (1998). The extension discussed in Sec. 3.9.2 is due to Grassi and Strini (1999).

The quantum search algorithm was introduced by Grover (1996), see also Grover (1997). The optimality of the quadratic speedup is discussed in Boyer *et al.* (1998) and Zalka (1999). Further developments can be found in Brassard *et al.* (2002).

Useful discussions on the quantum Fourier transform can be found in Coppersmith (1994) and Ekert and Jozsa (1996). The generalization of the quantum Fourier transform over a generic finite Abelian group was found by Kitaev (1995). Other useful references are Jozsa (1997), Ekert and Jozsa (1998) and Bowden *et al.* (2000). The quantum wavelet transform is discussed in Fijany and Williams (1998). A unified approach to fast unitary transforms is described in Agaian and Klappenecker (2002). Signal-processing methods in quantum computing are discussed in Klappenecker and Rötteler (2001).

The phase estimation algorithm was introduced by Kitaev (1995) and a good description of this algorithm can be found in Cleve *et al.* (1998).

The quantum algorithm described in Sec. 3.13, which computes eigenvalues and eigenvectors of a given unitary operator, was introduced by Abrams and Lloyd (1999).

Shor's algorithm for integer factoring was proposed in Shor (1994) and a detailed discussion of this algorithm can be found in Shor (1997). Other useful references are Kitaev (1995), Ekert and Jozsa (1996), Lomonaco (2000) and Beauregard (2003). A readable introduction to Shor's algorithm is Lavor *et al.* (2003).

The idea that a quantum computer might outperform a classical computer in simulating quantum mechanical systems was proposed by Feynman (1982) and further developed by Lloyd (1996). The method to simulate the Schrödinger equation in Sec. 3.15 is due to Zalka (1998) and Wiesner (1996). The study of quantum algorithms for the simulation of quantum chaos was started by Schack (1998) and further developed by Georgeot and Shepelyansky (2001a). The quantum algorithm for the simulation of the quantum sawtooth map is due to Benenti *et al.* (2001) and quantum computation of dynamical localization in this model is studied in Benenti *et al.* (2003). Other quantum algorithms for computing interesting physical quantities in dynamical models can be found in Emerson *et al.* (2002, 2003). A quantum circuit for quantum-state tomography is discussed in Miquel *et al.* (2002a). The quantum simulation of classical chaotic systems is discussed by Georgeot and Shepelyansky (2001b). The simulation of many-body Fermi systems on a quantum computer is investigated by Abrams and Lloyd (1997) and Ortiz *et al.* (2001). The simulation of spin systems is discussed in Sørensen and Mølmer (1999). The problem of simulating the equilibration of quantum systems on a quantum computer is addressed in Terhal and DiVincenzo (2000).

General references on quantum chaos are Casati and Chirikov (1995) and Haake (2000).

An interesting discussion of the criteria that must be fulfilled in order to realize a quantum computer can be found in DiVincenzo (2000). A review on decoherence is Zurek (2003) while a simple introduction can be found in Zurek (1991). There are several studies of the effect of decoherence and imperfections on the stability of quantum computation: for instance, Palma *et al.* (1996), Miquel *et al.* (1996), Georgeot and Shepelyansky (2000), Benenti *et al.* (2001) and Strini (2002).

A description of the basic principles of NMR quantum computation is given by Jones (2001). Various quantum algorithms have been implemented by means of few-qubit NMR quantum processors: for instance, Grover's algorithm (Jones, 1998) with 2 qubits, the quantum Fourier transform (Weinstein *et al.*, 2001) and the baker's map (Weinstein *et al.*, 2002) with 3 qubits and Shor's algorithm (Vandersypen *et al.*, 2001) with 7 qubits.

A scheme for efficient quantum computation with linear optics is discussed by Knill *et al.* (2001).

An excellent review of cavity quantum electrodynamics experiments manipulating the entanglement of atoms and photons is Raimond *et al.* (2001).

The idea to use ion traps for quantum computation was proposed by Cirac and Zoller (1995). The first demonstration of the CNOT gate is due to Monroe et al. (1995). The multipartite entanglement of four ions is described in Sackett *et al.* (2000). The implementation of the Deutsch–Jozsa algorithm on an ion-trap quantum computer is described in Gulde *et al.* (2003). The proposal to use an array of independent ion traps for quantum computation is discussed in Cirac and Zoller (2001).

A proposal to realize quantum computation with quantum dots is described in Loss and DiVincenzo (1998). The idea to use spins in silicon semiconductors for quantum computation was proposed by Kane (1998).

Rabi oscillation in Josephson junctions was demonstrated in Nakamura *et al.* (1999), Vion *et al.* (2002) and Chiorescu *et al.* (2003). Conditional gate operation using superconducting charge qubits was recently implemented by Yamamoto *et al.* (2003).

Chapter 4

Quantum Communication

In this chapter, we show that the basic properties of quantum mechanics can be put to practical use in the transmission of information. The most spectacular application is in the field of cryptography, the art of secret communication. After a short overview of classical cryptography, we discuss the unique contribution of quantum mechanics to cryptography. Quantum cryptography enables two communicating parties, named Alice (the *sender*) and Bob (the *receiver*), to detect whether the transmitted message has been intercepted by Eve (an *eavesdropper*). This is a consequence of a basic property of quantum mechanics, the "no-cloning theorem": an unknown quantum state cannot be cloned. Then we illustrate two remarkable applications of quantum mechanics: dense coding and quantum teleportation. Dense coding uses entanglement to enhance the communication of classical information. If Alice and Bob share an entangled EPR pair of qubits, Alice can operate on her member of the pair and then send it to Bob: this single qubit carries two bits of classical information. Quantum teleportation again exploits entanglement and allows Alice to transmit a quantum bit to Bob by sending him only classical bits. Finally, we provide a quick overview of the experimental implementations of quantum communication protocols and of their perspectives. The material of this section will be revisited in more depth in Chap. 8.

4.1 Classical cryptography

The origins of cryptography go back to before Christ, since when the need for secret communication has become evermore important. A significant example is the *Cæsar cypher*, used more than 2000 years ago by Julius Cæsar during the Gallic war. Such a code uses an alphabet in which each letter is shifted by a fixed number of steps. If there are, for instance, $k = 26$ letters in the alphabet, then we have $k - 1 = 25$ possible codes. If we label the letters as $1, 2, \ldots, k$, then the code number j is obtained with $i \to i + j$ (mod k), for $i = 1, 2, \ldots, k$. We say that the sender Alice encrypts her *plain text* into a *cypher text*, using a *secret key*. In the case of the Cæsar cypher, the key is j. It is clear that there are $k - 1$ different codes, singled out by the number j (with $j = 1, \ldots, k - 1$). The case $j = k$ is not acceptable since the original letters are unchanged. This code was difficult to break in Cæsar's time, but of course it is easily breakable today. Note that Alice must first of all communicate the key to Bob over a *secure channel*, assumed inaccessible to the eavesdropper Eve. After this Alice sends the cypher text to Bob over an *insecure channel*.

An interesting variant of the Cæsar cypher replaces the k letters of the alphabet with one of their $k!$ possible permutations. However, even this code is easy to break nowadays, since the letters of an alphabet appear with different frequencies in a text. Thus, a simple statistical analysis of the cypher text is sufficient to break the code.

4.1.1   *The Vernam cypher*

The first unbreakable code, the *Vernam cypher*, was invented in 1917 by Gilbert Vernam, even though the mathematical proof of its unbreakability was achieved only more than thirty years later by Shannon. Vernam's protocol is the following.

(i) The plain text is written as a binary sequence of 0's and 1's.
(ii) The secret key is a completely random binary sequence of the same length as the plain text.
(iii) The cypher text is obtained by adding the secret key bitwise modulo 2 to the plain text.

If $\{p_1, p_2, \ldots, p_N\}$ denotes the plain text (with $p_1, p_2, \ldots, p_N$ binary digits) and $\{k_1, k_2, \ldots, k_N\}$ the secret key, then the cypher text $\{c_1, c_2, \ldots, c_N\}$ is obtained as follows:

$$c_i = p_i \oplus k_i \qquad (i = 1, 2, \ldots, N). \tag{4.1}$$

Let us consider a simple example:

$$
\begin{array}{ll}
001010011 & \text{plain text,} \\
100111010 & \text{secret key,} \\
101101001 & \text{cypher text.}
\end{array}
\qquad (4.2)
$$

The code is unbreakable, provided that the key is completely random, since in this case the cypher text is completely random, too. It gives no information whatsoever about the plain text. Since the secret key is shared by Alice and Bob, the latter can easily reconstruct the plain text. He simply adds the secret key bitwise modulo 2 to the cypher text:

$$
p_i = c_i \oplus k_i \qquad (i = 1, 2, \ldots, N). \qquad (4.3)
$$

We stress that the secret key must be used only once (the Vernam cypher is also known as the *one time pad*). If it is reused and the eavesdropper is able to observe two cypher texts, then their bitwise addition modulo two is equal to the bitwise addition modulo two of the corresponding plain texts. Since in the plain texts there are always redundancies (they are not random binary sequences), the code becomes breakable. Therefore, the main problem of cryptography is not the transmission of the cypher text but the distribution of the secret key. This distribution requires some kind of "trusted courier"; that is, the problem of the secrecy of communication is merely transferred to the problem of the secrecy of the key. The problem is that Eve could, at least in principle, find a way to read the key without leaving any trace of her action. Therefore, Alice and Bob can never be absolutely sure of the secrecy of the key. We shall see in Sec. 4.3 that quantum mechanics solves this problem, offering a unique way for secure *key distribution* and *key storage*.

The Vernam cypher requires the generation of a long random binary string (at least as long as the message to be transmitted), a non-trivial task in itself. Weaker cyphers using shorter keys are in principle breakable, but possibly hard to break. It is worth noting that the task of breaking sophisticated cyphers was between the motivations that stimulated the construction of electronic computers.

4.1.2 *The public-key cryptosystem*

Owing to the difficulty of supplying new random keys for every message, the Vernam cypher is nowadays used mainly for important diplomatic communications. For less delicate business, it is substituted by *public-key* cryptographic systems, whose principles were discovered in the middle of the 1970's by Diffie and Hellman.

The fundamental difference between the traditional secret-key cryptosystem and the recent public-key cryptosystem is the following.

(i) In the secret-key cryptosystem, Alice encrypts her message by means of a secret key. She sends the encrypted message to Bob, who owns the same secret key and can therefore decrypt the message. The security of the message resides in the secrecy of the key. Since the secret key must be at some time distributed between Alice and Bob, there is always a risk that the key is intercepted.

(ii) In the public-key cryptosystem, Alice and Bob do not exchange any secret key. Bob makes public a key (the public key), used by Alice to encrypt the message. However, the message cannot be decrypted by this key, but only by another key (the private key), which is possessed by Bob alone. Therefore, the key-distribution problem is avoided. The public-key cryptosystem works as if Bob had constructed a safe into which he inserted the message. The safe has two keys, one public to lock it and another private to open it. Anyone may place a message in the safe, but only one person (Bob) can open the safe and take the message out.

The public-key cryptosystem requires the use of a *trap-door function* f, easy to compute but with inverse function f^{-1} hard to compute. Here the words easy and hard must be understood according to the theory of computational complexity (see Chap. 1): f can be computed with resources polynomial in the input size, while f^{-1} cannot be computed with polynomial resources (resources denote computational time, size of the hardware *etc.*). Any problem whose solution is hard to find but easy to verify is in principle useful for cryptography. These problems lie in the computational class **NP**. Two keys are involved: a public key f, used by Alice to encrypt her text, and a secret key f^{-1}, possessed by Bob alone, who uses it to decrypt the message.

4.1.3 *The RSA protocol*

A famous example of public-key cryptosystem is the RSA cryptosystem, devised in 1977 by Rivest, Shamir and Adleman. The RSA protocol works as follows:

1. Bob chooses two "large enough" prime numbers p and q, and computes $pq = N$.
2. Bob chooses at random a number d that is co-prime with $(p-1)(q-1)$, that is, the greatest common divisor of d and $(p-1)(q-1)$ is equal to 1.
3. Bob computes e, the inverse modulo $(p-1)(q-1)$ of d:

$$e\,d|_{\mathrm{mod}(p-1)(q-1)} = 1. \tag{4.4}$$

Note that from now on we can forget about p and q.
4. Bob publishes the pair (e, N). This is the public key, which anybody can use to send messages to Bob.
5. The pair (d, N) is the private decryption key, possessed by Bob alone. Therefore, only Bob can decrypt the messages that were encrypted by means of the public key.
6. Alice divides her message into blocks and each block can be written as a number. For the i-th block, the number is m_i, with $m_i < N$. Alice encrypts each block as follows:

$$m_i \to m'_i = m_i^e|_{\mathrm{mod}\,N}. \tag{4.5}$$

7. Bob decrypts the message by computing

$$m_i = m_i'^{\,d}|_{\mathrm{mod}\,N}. \tag{4.6}$$

Indeed, elementary number theory tells us that $m_i^{ed}|_{\mathrm{mod}\,N} = m_i$ (see, *e.g.*, Ekert *et al.*, 2001).

Note the advantages with respect to the Vernam cypher:

(i) there is no need to distribute a secret key over a supposedly secure channel: the public key can be used by anybody who wishes to communicate with Bob while the secret key is possessed by Bob alone;
(ii) the public key can be reused as many times as desired.

Exercise 4.1 A crucial problem of cryptography is *authentication*: Bob needs to determine if the message received was really sent by Alice and not by someone else. Find how Alice can authenticate her message using a public-key cryptographic system.

The RSA code can be broken if one discovers the prime factors p and q of N. After this d can be easily computed since e is known. Therefore, the reliability of the method is based on the fact that there exist no known efficient (polynomial time) algorithms to find the factors of an integer N: The best classical algorithm known today for integer factorization, the number field sieve, requires $\exp(O(n^{1/3}(\log n)^{2/3}))$ operations, where $n = \log N$ is the input size. To give a concrete idea of the difficulty of the problem, the factorization of a number 250 digits long would take about 10 million years on a 200-MIPS (million of instructions per second) computer (see Hughes, 1998). This means that the problem is in practice impossible to solve with current technology. To be more precise, we should add that it is not proved that the number field sieve algorithm is optimal for integer factorization. Furthermore, the possibility that a polynomial time algorithm can be discovered is not excluded, just as it cannot be excluded for any other **NP** problem. In any event, we should like to stress that, as we have seen in Sec. 3.14, such a polynomial time algorithm exists on a quantum computer. Therefore, a large-scale quantum computer, if constructed, would break the RSA encryption scheme. This is a clear demonstration that the security of public-key cryptosystems is not a sufficient guarantee for messages that must be kept secret for indefinitely long times.

4.2   The no-cloning theorem

There is one property of the classical bit that we take for granted: it can be copied. In contrast, we shall see in this section that the *generic* state of a qubit cannot be cloned. This is the content of the so-called no-cloning theorem of Dieks, Wootters and Zurek.

Let us first consider a concrete example, in which the qubit is the polarization state of a photon. We label the state of a photon as $|0\rangle$ (or $|\leftrightarrow\rangle$) when it is in a horizontally polarized state and as $|1\rangle$ (or $|\updownarrow\rangle$) when it is in a vertically polarized state. A photon can also be polarized along a direction forming an angle β with respect to the horizontal (see Fig. 4.1, where x denotes the horizontal axis, y the vertical axis and z the direction of propagation of the photon). In this case, it is described by the wave function

$$|\psi\rangle \,=\, \cos\beta\,|\leftrightarrow\rangle + \sin\beta\,|\updownarrow\rangle. \tag{4.7}$$

Now assume that such a photon is sent to a polarization analyzer (a birefringent crystal such as calcite). The photon emerges from the analyzer horizontally polarized or vertically polarized (see Fig. 4.1, where horizontally polarized photons pass straight through the crystal, while vertically polarized photons are deflected). These two mutually exclusive outcomes, which we call 0 and 1, take place with probabilities $p_0 = |\langle \leftrightarrow |\psi\rangle|^2 = \cos^2 \beta$ and $p_1 = |\langle \updownarrow |\psi\rangle|^2 = \sin^2 \beta$, respectively. Therefore, a measurement of the polarization state of a single photon gives a single bit of information, corresponding to the polarization state of the detected photon. We note that this is in perfect agreement with the measurement postulate introduced in Sec. 2.4 (Postulate II of quantum mechanics).

Assume now that a cloning machine exists. Then one could make an arbitrarily large number of copies of the state $|\psi\rangle$ given by Eq. (4.7). Thus, it would be possible to measure all these clones and obtain the angle β to any desired accuracy. Since the cloning machine can be thought of as part of the measurement apparatus, this would contradict the measurement postulate. This postulate implies that from the measurement of the polarization state of a photon we can obtain only a single bit of information. We obtain 0 with probability $p_0 = \cos^2 \beta$ and 1 with probability $p_1 = \sin^2 \beta$. In contrast, if a cloning machine existed, from this measurement we could determine, to any desired accuracy, the parameter β. Therefore, from the simple measurement of the polarization state of a single photon one could extract an arbitrarily large amount of information (the bits necessary to represent β to the desired accuracy). We can conclude, on the basis of the measurement postulate of quantum mechanics, that a quantum cloning machine cannot exist.

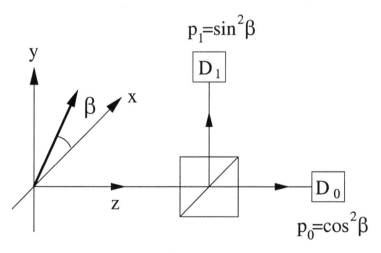

Fig. 4.1 Measurement of the polarization of a single photon. The two photodetectors are denoted by D_0 and D_1.

We now give a more formal proof of the no-cloning theorem. This proof has to be considered weaker than the previous, since it takes advantage of the linearity of quantum mechanics.

Theorem 4.1 *It is impossible to build a machine that operates unitary transformations and is able to clone the generic state of a qubit.*

Proof. Let us consider a system composed of the qubit to be cloned, a second qubit and the cloning machine. The first qubit is prepared in a generic state

$$|\psi\rangle = \alpha|0\rangle + \beta|1\rangle, \tag{4.8}$$

with α and β complex numbers, constrained by the normalization condition $|\alpha|^2 + |\beta|^2 = 1$. Initially, the second qubit and the cloning machine are prepared in some reference states, for instance $|\phi\rangle$ and $|A_i\rangle$, respectively. The cloning machine should be able to perform a unitary transformation U such that

$$U\big(|\psi\rangle|\phi\rangle|A_i\rangle\big) = |\psi\rangle|\psi\rangle|A_{f\psi}\rangle = \big(\alpha|0\rangle + \beta|1\rangle\big)\big(\alpha|0\rangle + \beta|1\rangle\big)|A_{f\psi}\rangle, \tag{4.9}$$

where the final state of the machine will in general depend on the state $|\psi\rangle$ to be cloned. We show now that such a unitary transformation cannot exist. If the first qubit is in the state $|0\rangle$, the action of the cloning machine must be

$$U\big(|0\rangle|\phi\rangle|A_i\rangle\big) = |0\rangle|0\rangle|A_{f0}\rangle. \tag{4.10a}$$

Analogously, if the first qubit is prepared in the state $|1\rangle$,

$$U\big(|1\rangle|\phi\rangle|A_i\rangle\big) = |1\rangle|1\rangle|A_{f1}\rangle. \tag{4.10b}$$

Therefore, the action of the cloning machine on a generic state $|\psi\rangle = \alpha|0\rangle + \beta|1\rangle$ is given by

$$U\big((\alpha|0\rangle + \beta|1\rangle)|\phi\rangle|A_i\rangle\big) = \alpha\, U\big(|0\rangle|\phi\rangle|A_i\rangle\big) + \beta\, U\big(|1\rangle|\phi\rangle|A_i\rangle\big), \tag{4.11}$$

where we have invoked the *linearity* of quantum mechanics. We now insert Eqs. (4.10a) and (4.10b) into Eq. (4.11), obtaining the state

$$\alpha\,|0\rangle|0\rangle|A_{f0}\rangle + \beta\,|1\rangle|1\rangle|A_{f1}\rangle, \tag{4.12}$$

which is clearly different from the desired cloned state of Eq. (4.9). $\square$

We stress that it is essential to consider a *generic* state. Indeed, if we know from the beginning that the quantum state of the qubit is prepared in one out of two orthogonal states, for instance $|0\rangle$ or $|1\rangle$, then we can measure with certainty the state of the qubit and prepare as many copies of it as desired. In this case the qubit acts as a classical bit and we know that there exist classical cloning machines (photocopiers *etc.*).

Exercise 4.2 A single qubit is in an unknown state $|\psi_1\rangle$. We guess at random that its state is $|\psi_2\rangle$. What is the average fidelity F of our guess, defined by $F \equiv \left|\langle\psi_1|\psi_2\rangle\right|^2$?

4.2.1 *Faster-than-light transmission of information?*

It is interesting to show that the existence of a cloner would violate a basic principle of the theory of relativity, that is, information could be transmitted faster than light. Assume that a source produces EPR pairs in the Bell state

$$|\phi^+\rangle = \tfrac{1}{\sqrt{2}}\left(|0\rangle|0\rangle + |1\rangle|1\rangle\right). \tag{4.13a}$$

This state can also be written as

$$|\phi^+\rangle = \tfrac{1}{\sqrt{2}}\left(|0\rangle_x|0\rangle_x + |1\rangle_x|1\rangle_x\right), \tag{4.13b}$$

where $|0\rangle_x$ and $|1\rangle_x$ are the eigenstates of the Pauli matrix σ_x with eigenvalues $+1$ and -1, respectively. It is easy to check the equality of expressions (4.13a) and (4.13b), taking into account that

$$\begin{cases} |0\rangle_x = \tfrac{1}{\sqrt{2}}\begin{bmatrix} 1 \\ 1 \end{bmatrix} = \tfrac{1}{\sqrt{2}}\left(|0\rangle + |1\rangle\right), \\[2mm] |1\rangle_x = \tfrac{1}{\sqrt{2}}\begin{bmatrix} 1 \\ -1 \end{bmatrix} = \tfrac{1}{\sqrt{2}}\left(|0\rangle - |1\rangle\right). \end{cases} \tag{4.14}$$

A member of each EPR pair is sent to Alice, the other to Bob. Note that, in principle, Alice and Bob can be located arbitrarily far apart. Alice codes a message that she wants to send to Bob in a binary string. She then performs a measurement on her member of each EPR pair and she chooses to measure σ_x or σ_z depending on the fact that the corresponding digit is 0 or 1 (we assume that Alice and Bob share at least as many EPR pairs as digits in their message). After this, the state of Bob's member of the EPR pair collapses onto an eigenstate of σ_x or σ_z. However, these states are not orthogonal and Bob cannot obtain any information about Alice's message from his measurements. In contrast, if a cloning machine existed, Bob could make an arbitrarily large number of copies of his EPR qubits and distinguish between eigenstates of σ_x and σ_z with any desired accuracy. Indeed, given the generic state of a qubit, $|\psi\rangle = \alpha|0\rangle + \beta|1\rangle$, and a large number of copies of this qubit, Bob could estimate α and β from measuring these qubits. Hence, in particular, he could determine whether this state were an eigenstate of σ_x or of σ_z. Therefore, superluminal transmission of information would be possible, in contradiction with a basic principle of the theory of relativity.

Exercise 4.3   Show that, in the above example, independently of the decision of Alice to measure σ_x or σ_z, Bob obtains an up or down polarization state with equal probability, whatever direction he chooses for the measurement of the spin polarization.

Finally, we note that the no-cloning theorem does not forbid the construction of *imperfect* copies of the original state. Many schemes have been proposed that optimize some measure of the fidelity of the copies. We shall illustrate a concrete example of an imperfect quantum cloning machine in Chap. 5.

4.3   Quantum cryptography

In classical physics it is impossible to know with certainty if the eavesdropper Eve is monitoring a message. The reason is that classical information can be copied without changing the original message. Indeed, information must be encoded in some physical system (a piece of paper, radio signals *etc.*), whose properties can, in principle, be measured *passively*. The modifications induced in the system can be made as small as permitted by the available technology. In contrast, in quantum mechanics the measurement process in general disturbs the system for fundamental reasons. This is a consequence of the Heisenberg uncertainty principle (see Sec. 2.4). Indeed, if one considers a pair of non-commuting observables, the measurement of one observable necessarily disturbs (randomizes) the other. In this section, we shall see that this inherently quantum property allows *intrusion detection*: Alice and Bob can discover if Eve is eavesdropping their communication. This possibility can be used to create a secret key between two parties. The resulting key allows Alice and Bob to communicate secretly by means of classical cryptosystems like the Vernam cypher. In the following we describe two protocols for quantum-key distribution, the BB84 protocol and the E91 protocol.

4.3.1 *The BB84 protocol*

The protocol BB84, discovered by Bennett and Brassard in 1984, requires four states and two binary alphabets: $|0\rangle$ and $|1\rangle$ (the z-alphabet), $|+\rangle \equiv |0\rangle_x = \frac{1}{\sqrt{2}}(|0\rangle + |1\rangle)$, $|-\rangle \equiv |1\rangle_x = \frac{1}{\sqrt{2}}(|0\rangle - |1\rangle)$ (the x-alphabet). The letters of the z- and x-alphabets are associated with the eigenstates of the Pauli matrices σ_z and σ_x, respectively. The description of the BB84 protocol follows (and a simple example is given in Table 4.1):

1. Alice generates a random sequence of 0's and 1's.
2. Alice encodes each data bit in a qubit, $|0\rangle$ or $|+\rangle = |0\rangle_x$ if the corresponding bit is 0, $|1\rangle$ or $|-\rangle = |1\rangle_x$ if the corresponding bit is 1. For each bit, Alice chooses randomly between the x- and the z-alphabet, by means of a fair coin (*e.g.*, if the coin lands heads Alice chooses the x-alphabet, while the z-alphabet is chosen when the coin lands tails).
3. The resulting string of qubits is sent by Alice and received by Bob.
4. For each qubit, Bob decides at random what axis (alphabet) to use for the measurement, x or z. In the first case, he measures the spin polarization along the x axis, in the latter along the z axis. Note that half of the time Bob chooses the same axis as Alice. In this case, assuming that there are no eavesdroppers or noise effects, Alice and Bob share the same bit (here we summarize under the word noise effects such as imperfect state preparation or detection, interactions of the transmitted qubit with the environment *etc.*). In contrast, if Bob chooses an axis different from Alice, the bit resulting from his measurement agrees with the bit sent by Alice only half of the time. For instance, if Bob receives the qubit $|-\rangle$ and measures σ_z, the outcomes 0 and 1 have equal probability.
   From now on Alice and Bob exchange only classical information over a public channel.
5. Bob communicates to Alice over a classical public channel which alphabet he used for each qubit measurement. Of course, he does not communicate the results of these measurements.
6. Alice communicates to Bob over a classical public channel which alphabet she used for each transmitted qubit (again, not the results of these measurements).
7. Alice and Bob delete all bits corresponding to the cases in which they

used different alphabets. After this they share the so-called *raw key*. This key is the same for Alice and Bob, insofar as Eve and noise were absent.

By means of the following steps, Alice and Bob distill the secret key starting from the raw key.

8. Over a public communication channel, Alice and Bob announce and compare a part of their raw key. From this comparison they can estimate the *error rate R* due to eavesdroppers or noise effects. If this rate is too high they restart the protocol from the beginning. If not, they perform *information reconciliation* and *privacy amplification* on the remaining bits of their raw key.

9. Information reconciliation is just classical error correction over a public transmission channel. We shall describe classical error-correcting codes in Chap. 7. Here, we limit ourselves to the illustration of a simple scheme for information reconciliation. Alice and Bob divide the remaining bits of their raw key into subsets of length l. This length is chosen in order that it is unlikely to have more than one error per subset ($Rl \ll 1$, with R the previously estimated error rate). For each subset, Alice and Bob make parity checks (the parity P of a binary string $\{b_1, b_2, \ldots, b_l\}$ is defined as $P = b_1 \oplus b_2 \oplus \cdots \oplus b_l$), discarding each time the last bit. If the parities of a given subset are different for Alice and Bob, then they locate and delete the erroneous bit by binary search in the following way. They bisect the subset and check the parities of the new blocks ($P_1 = b_1 \oplus b_2 \oplus \cdots \oplus b_{(l-1)/2}$ and $P_2 = b_{(l-1)/2+1} \oplus b_{(l-1)/2+2} \oplus \cdots \oplus b_{l-1}$). They repeat the bisection for the block in which Alice's and Bob's parities differ and so on. Note that each time Alice and Bob delete the last bit of the blocks whose parity is publicly announced. In this way, they avoid Eve obtaining any amount of information from their parity checks. At the end, with high probability, Alice and Bob share the same string of bits.

10. Privacy amplification reduces Eve's information about the final secret key to arbitrarily small values. Let us illustrate a simple privacy amplification protocol. Alice and Bob estimate from the error rate R obtained previously the maximum number of bits k known by Eve. Let s be a security parameter. Then Alice and Bob choose at random $n - k - s$ subsets of their key, where n denotes the number of bits in the key. The parities of these subsets become the final secret key. This key is more secure than the previous one, since Eve must know something about each bit of a subset in order to obtain information about its parity. It can be shown that Eve's residual information is $O(2^{-s})$.

Note that Eve can choose different eavesdropping strategies:

(i) *intercept and resend*: Eve intercepts and measures the qubits sent by Alice and resends them to Bob;

(ii) *translucent attack*: Eve has probes (ancillary qubits) interacting with the qubits sent by Alice and she measures the state of these probes;

(iii) *collective attack*: Eve manipulates not a single qubit at a time but a block of qubits.

Independently of the eavesdropping strategy, it is possible to show that quantum-key distribution is secure, in the sense that it is possible to guarantee that Eve's information about the final key is arbitrarily small (see Nielsen and Chuang, 2000).

Table 4.1 An example of the BB84 protocol.

Alice's data bits	1	0	0	0	1	1	0	1	0	1
Alice's alphabet	x	z	x	z	x	x	x	z	z	x
Transmitted qubits	$|1\rangle_x$	$|0\rangle$	$|0\rangle_x$	$|0\rangle$	$|1\rangle_x$	$|1\rangle_x$	$|0\rangle_x$	$|1\rangle$	$|0\rangle$	$|1\rangle_x$
Bob's alphabet	x	z	x	x	z	x	z	x	z	z
Measurement outcomes	1	0	0	0	0	1	0	0	0	1
Bob's data bits	1	0	0	0	0	1	0	0	0	1
Raw key	1	0	0			1			0	

We stress that the validity of the BB84 protocol is based on the Heisenberg principle. The two alphabets are associated with two non-commuting observables, σ_x and σ_z. Eve cannot measure both the polarization along x and along z for the same qubit. For instance, if she measures σ_z for the qubit $|0\rangle_x$, she obtains the outcomes 0 or 1 with equal probability. Thus, she has irreversibly randomized the polarization originally sent by Alice. We also stress the importance of the no-cloning theorem: it guarantees that Eve cannot distinguish with certainty between non-orthogonal quantum states. If a quantum cloning machine existed, Eve could make a large number of copies of each qubit sent by Alice and distinguish with arbitrary accuracy between eigenstates of σ_x and σ_z. For instance, assume that Eve measures σ_z for the qubit and all its copies. If she received $|1\rangle$, she always obtains outcome 1. On the other, if she received $|1\rangle_x$, she obtains outcomes 0 and 1 with equal probabilities. Finally, Eve could resend a copy of the intercepted qubit to Bob. Therefore, if it were possible to violate the no-cloning theorem, Eve could intercept the qubits sent by Alice and resend them to Bob, leaving no trace of her intrusion.

Exercise 4.4   Show that if Eve intercepts every qubit sent by Alice, measures its polarization along some axis and resends it to Bob, then she introduces an error rate of 25% in the raw key.

Exercise 4.5   Show that it is not possible to gain information about which one of two non-orthogonal quantum states $|\psi_1\rangle$ and $|\psi_2\rangle$ was sent by Alice without disturbing the state.

Finally, we note that one of the main drawbacks of quantum cryptography is that no mechanism is known for authentication. Thus, a classical secret key is required for this purpose. Indeed, in order to be sure that they are not communicating with someone else, Alice and Bob need to send an authentication key over a classical secure channel. After this they can implement a quantum protocol like BB84 and "expand" the existing authentication key.

4.3.2   *The E91 protocol*

Now we discuss the protocol E91 (Ekert, 1991), a quantum cryptosystem that uses entangled EPR pairs.

1. A source S emits a pair of qubits (spin $\frac{1}{2}$ particles) in the EPR state

$$|\psi^-\rangle = \tfrac{1}{\sqrt{2}}\left(|01\rangle - |10\rangle\right). \qquad (4.15)$$

   The qubits are sent in opposite directions; the first is received by Alice, the second by Bob (see Fig. 4.2). Note that a third party is not strictly necessary: Alice could produce EPR pairs and then send a member of each pair to Bob.

2. Alice and Bob can discover if Eve has intercepted the transmission of the EPR pairs by exploiting the quantum correlations of EPR pairs. They measure the spin polarization along one of the three directions $\hat{a}_1$, $\hat{a}_2$, $\hat{a}_3$ (Alice) and $\hat{b}_1$, $\hat{b}_2$, $\hat{b}_3$ (Bob) (see Fig. 4.3). For each EPR pair, Alice and Bob choose at random their measurement axes between $\hat{a}_1$, $\hat{a}_2$, $\hat{a}_3$ and $\hat{b}_1$, $\hat{b}_2$, $\hat{b}_3$, respectively. Let us denote $p_{\pm\pm}(\hat{a}_i, \hat{b}_j)$ the probability that Alice's polarization measurement along the direction $\hat{a}_i$ gives the result ± 1 and Bob's measurement along $\hat{b}_i$ gives ± 1. We define the correlation

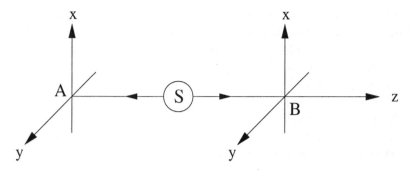

Fig. 4.2 A schematic picture of the E91 protocol. The EPR source, Alice, and Bob are denoted by S, A and B.

coefficients:

$$E(\hat{a}_i, \hat{b}_j) = p_{++}(\hat{a}_i, \hat{b}_j) + p_{--}(\hat{a}_i, \hat{b}_j) - p_{+-}(\hat{a}_i, \hat{b}_j) - p_{-+}(\hat{a}_i, \hat{b}_j). \quad (4.16)$$

From the discussion of Sec. 2.5 on Bell's inequalities, we know that

$$C \equiv E(\hat{a}_1, \hat{b}_1) - E(\hat{a}_1, \hat{b}_3) + E(\hat{a}_3, \hat{b}_1) + E(\hat{a}_3, \hat{b}_3) = -2\sqrt{2}, \quad (4.17)$$

that is, quantum mechanics violates the CHSH inequality, which reads $|C| \leq 2$ (see Sec. 2.5).

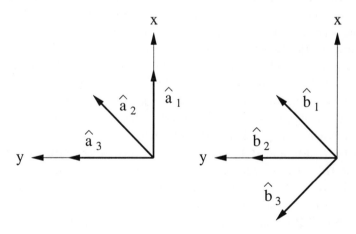

Fig. 4.3 Directions of the measurement axes for Alice (left) and Bob (right). The angles between these directions and the x axis are 0, $\frac{\pi}{4}$, $\frac{\pi}{2}$ for $\hat{a}_1$, $\hat{a}_2$, $\hat{a}_3$ and $\frac{\pi}{4}$, $\frac{\pi}{2}$, $\frac{3\pi}{4}$ for $\hat{b}_1$, $\hat{b}_2$, $\hat{b}_3$.

3. Alice and Bob announce over a public channel the axis chosen for each measurement. Then they make public the outcomes of the measurements in the cases in which their polarization axes did not coincide. This allows Alice and Bob to check the equality (4.17). If $C > -2\sqrt{2}$, then Eve attacked the EPR pairs or there were noise effects (note that it is possible to show that $|C|$ cannot be larger than $2\sqrt{2}$; for a proof of this result see, *e.g.*, Preskill, 1998). In the absence of such effects, that is, $C = -2\sqrt{2}$, Alice and Bob's measurements along the same axis are perfectly anticorrelated, namely,

$$E(\hat{a}_2, \hat{b}_1) = E(\hat{a}_3, \hat{b}_2) = -1. \qquad (4.18)$$

The outcomes of these measurements are the raw key shared by Alice and Bob (the keys agree if Bob negates his outcomes, $0 \to 1$ and $1 \to 0$). After this Alice and Bob can perform key reconciliation and privacy amplification as in the BB84 protocol.

We note that it is not necessary to test the relation (4.17). Alice and Bob could simply perform measurements along x or z. The decision is random, each choice occurring with probability $\frac{1}{2}$. After the measurement process, Alice and Bob announce over a public channel which observable they measured for each EPR pair. In the cases in which their measurement axes agree, the outcomes are perfectly anticorrelated. Alice and Bob discard the other bits, thus remaining with a shared raw key. After this, they proceed as in the BB84 protocol. It is interesting to note that the secret key is not generated by either Alice or Bob. The key is undetermined until Alice and Bob measure their halves of the shared EPR pairs. Then the secret key arises from a fundamentally random process, the quantum measurement. Finally, we stress that the E91 protocol is potentially interesting for *key storage*. The problem is the following: once the secret key has been established, Alice and Bob must store it in their safes, until they need it. However, the key is a string of *classical bits* and, in principle, can be copied. It may be very difficult to crack the safe, but always possible. No fundamental reasons exclude this possibility. However, if Alice and Bob were able to store EPR pairs, they could wait to establish the secret key until needed. Of course, the implementation of such key storage is hampered by the fact that one should be able to protect the EPR pairs from noise effects, due to interactions with their environment, for long times. This possibility is beyond the reach of present technology.

4.4 Dense coding

Dense coding is the simplest example of the application of quantum entanglement to communication. It allows Alice to send two bits of classical information to Bob by sending him only a single qubit. The dense coding protocol works as follows (see the schematic picture in Fig. 4.4 and the quantum circuit implementing the protocol in Fig. 4.5):

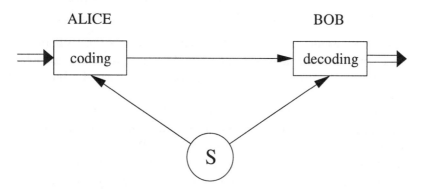

Fig. 4.4 A schematic picture of the dense coding protocol. The double lines denote two classical bits and the single lines a quantum bit.

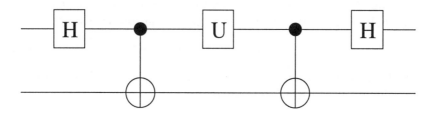

Fig. 4.5 A quantum circuit implementing the dense coding protocol.

1. A source S generates an EPR pair shared by Alice and Bob. The EPR pair is prepared, for instance, in the state

$$|\phi^+\rangle = \tfrac{1}{\sqrt{2}}(|00\rangle + |11\rangle). \qquad (4.19)$$

The Bell state $|\phi^+\rangle$ is obtained from the state $|00\rangle$ after application of a Hadamard gate and a CNOT gate:

$$\text{CNOT}\,(H \otimes I)\,|00\rangle = |\phi^+\rangle. \qquad (4.20)$$

In the computational basis with basis vectors $|00\rangle$, $|01\rangle$, $|10\rangle$ and $|11\rangle$ (note that the first label refers to Alice's half of the EPR pair, the second to Bob's half), the transformation (4.20) has the following matrix representation:

$$
\begin{bmatrix} 1 & 0 & 0 & 0 \\ 0 & 1 & 0 & 0 \\ 0 & 0 & 0 & 1 \\ 0 & 0 & 1 & 0 \end{bmatrix} \frac{1}{\sqrt{2}} \begin{bmatrix} 1 & 0 & 1 & 0 \\ 0 & 1 & 0 & 1 \\ 1 & 0 & -1 & 0 \\ 0 & 1 & 0 & -1 \end{bmatrix} \begin{bmatrix} 1 \\ 0 \\ 0 \\ 0 \end{bmatrix} = \frac{1}{\sqrt{2}} \begin{bmatrix} 1 \\ 0 \\ 0 \\ 1 \end{bmatrix} . \tag{4.21}
$$

Note that, as shown in Fig. 4.4, a source S creates the EPR pair and then sends one member of the pair to Alice and the other to Bob. It is important to stress that Alice and Bob can be located arbitrarily far apart.

2. There are four possible values of the two classical bits that Alice wishes to send to Bob: 00, 01, 10 and 11. They determine the unitary operation U that Alice performs on her half of the EPR pair: $U = I$, σ_x, σ_z or $i\sigma_y$ (we remind the reader that I denotes the identity and σ_x, σ_y, σ_z the Pauli matrices). The reason for choosing one of these four unitary transformations will become clear in the following.

As we have said, the operator U in the circuit of Fig. 4.5 is determined by the value of the two classical bits that Alice wishes to communicate to Bob. If she wishes to communicate 00, she operates the identity on her half of the EPR pair. This gives the trivial transformation

$$
I \otimes I \, |\phi^+\rangle = |\phi^+\rangle . \tag{4.22}
$$

If she wishes to communicate 01, she operates the Pauli matrix σ_x on her half of the EPR pair, obtaining

$$
\sigma_x \otimes I \, |\phi^+\rangle = |\psi^+\rangle , \tag{4.23}
$$

which corresponds to the following matrix representation in the computational basis:

$$
\begin{bmatrix} 0 & 0 & 1 & 0 \\ 0 & 0 & 0 & 1 \\ 1 & 0 & 0 & 0 \\ 0 & 1 & 0 & 0 \end{bmatrix} \frac{1}{\sqrt{2}} \begin{bmatrix} 1 \\ 0 \\ 0 \\ 1 \end{bmatrix} = \frac{1}{\sqrt{2}} \begin{bmatrix} 0 \\ 1 \\ 1 \\ 0 \end{bmatrix} . \tag{4.24}
$$

If she wishes to communicate 10, she operates σ_z, obtaining

$$
\sigma_z \otimes I \, |\phi^+\rangle = |\phi^-\rangle , \tag{4.25}
$$

with matrix representation

$$
\begin{bmatrix}
1 & 0 & 0 & 0 \\
0 & 1 & 0 & 0 \\
0 & 0 & -1 & 0 \\
0 & 0 & 0 & -1
\end{bmatrix}
\frac{1}{\sqrt{2}}
\begin{bmatrix}
1 \\ 0 \\ 0 \\ 1
\end{bmatrix}
=
\frac{1}{\sqrt{2}}
\begin{bmatrix}
1 \\ 0 \\ 0 \\ -1
\end{bmatrix}.
\tag{4.26}
$$

Finally, if she wishes to communicate 11, she operates $i\sigma_y$, obtaining

$$
i\sigma_y \otimes I |\phi^+\rangle = |\psi^-\rangle,
\tag{4.27}
$$

with matrix representation

$$
\begin{bmatrix}
0 & 0 & 1 & 0 \\
0 & 0 & 0 & 1 \\
-1 & 0 & 0 & 0 \\
0 & -1 & 0 & 0
\end{bmatrix}
\frac{1}{\sqrt{2}}
\begin{bmatrix}
1 \\ 0 \\ 0 \\ 1
\end{bmatrix}
=
\frac{1}{\sqrt{2}}
\begin{bmatrix}
0 \\ 1 \\ -1 \\ 0
\end{bmatrix}.
\tag{4.28}
$$

Up to this point, the circuit in Fig. 4.5 has constructed one of the four Bell states defined in Sec. 3.4 ($|\phi^+\rangle$, $|\psi^+\rangle$, $|\phi^-\rangle$ and $|\psi^-\rangle$).

3. Alice sends her half of the EPR pair to Bob.
4. Bob performs the appropriate unitary operations on the EPR pair, measures the two qubits and obtains the two classical bits. First of all, Bob transforms the Bell states into states of the computational basis. As was described in Sec. 3.4, the appropriate circuit for this operation is the inverse of the circuit in Fig. 3.7, which is also the first part of the dense-coding circuit represented in Fig. 4.5. Since Hadamard and CNOT gates are self-inverse, Bob operates

$$
\left(\text{CNOT} \, (H \otimes I) \right)^{-1} = (H \otimes I) \, \text{CNOT},
\tag{4.29}
$$

which has matrix representation

$$
B = \frac{1}{\sqrt{2}}
\begin{bmatrix}
1 & 0 & 0 & 1 \\
0 & 1 & 1 & 0 \\
1 & 0 & 0 & -1 \\
0 & 1 & -1 & 0
\end{bmatrix}.
\tag{4.30}
$$

It is easy to check that

$$
\begin{array}{ll}
B|\phi^+\rangle = |00\rangle, & \qquad B|\psi^+\rangle = |01\rangle, \\
B|\phi^-\rangle = |10\rangle, & \qquad B|\psi^-\rangle = |11\rangle.
\end{array}
\tag{4.31}
$$

Eventually, Bob measures the two qubits in the computational basis, obtaining with unit probability the two desired classical bits.

We stress that dense coding is not possible in classical physics, since a classical bit also has a well-defined value prior to its measurement. In quantum mechanics, there is entanglement. When Alice operates on her half of the EPR pair, she acts not on an isolated qubit, but on an entangled two-qubit system.

4.5 Quantum teleportation

Quantum teleportation is one of the most amazing applications of quantum physics to the realm of information theory: it allows for the transmission of quantum information from Alice to Bob, even though Alice sends only classical information to Bob. This possibility could be of practical interest for quantum computation, for example in the transfer of quantum information between different units of a quantum computer. Let us consider the simplest example of teleportation: Alice owns a two level system in some unknown state

$$|\psi\rangle = \alpha|0\rangle + \beta|1\rangle, \tag{4.32}$$

and she wishes to send this qubit to Bob using only a classical communication channel: she can send only classical bits, not quantum bits. At first sight, the task seems desperate since a measurement of the system would uncontrollably perturb its state and from this measurement Alice can obtain only a single bit of information. We note that describing $|\psi\rangle$ requires an infinite amount of classical information, since this quantum state lives in a continuous space (it is parametrized by two complex parameters α and β). However, quantum teleportation solves the problem, provided that Alice and Bob share an entangled pair of qubits. The protocol for quantum teleportation is outlined in the following (the quantum circuit implementing teleportation is shown in Fig. 4.6):

1. The first two gates of the circuit in Fig. 4.6 create the Bell state

$$|\psi^+\rangle = \tfrac{1}{\sqrt{2}}\big(|01\rangle + |10\rangle\big). \tag{4.33}$$

Indeed,

$$\mathrm{CNOT}\,(H \otimes I)\,|01\rangle = |\psi^+\rangle. \tag{4.34}$$

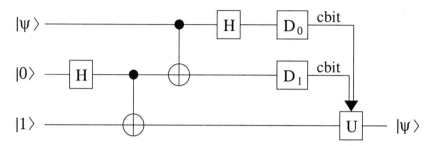

Fig. 4.6 A quantum circuit for teleportation. The first line represents the qubit to be teleported, the second line a qubit possessed by Alice, and the third a qubit possessed by Bob. The measurement performed by Alice (by means of the detectors D_0 and D_1) gives two *cbits* (classical bits) of information, which control the unitary transformation U performed by Bob.

They are operated by a source S that generates the EPR pair. Then the first half of the EPR pair is sent to Alice and the second half to Bob. Therefore, Alice owns two qubits (the state $|\psi\rangle$ and half of the EPR pair) and Bob a single qubit (the second half of the EPR pair). Note that, as usual, Alice and Bob can be very far apart. The three-qubit state is given by the direct product

$$
\begin{aligned}
|\psi\rangle \otimes |\psi^+\rangle &= \left(\alpha|0\rangle + \beta|1\rangle \right) \otimes \tfrac{1}{\sqrt{2}} \left(|01\rangle + |10\rangle \right) \\
&= \tfrac{\alpha}{\sqrt{2}} \left(|001\rangle + |010\rangle \right) + \tfrac{\beta}{\sqrt{2}} \left(|101\rangle + |110\rangle \right).
\end{aligned} \qquad (4.35)
$$

2. Alice allows the qubit $|\psi\rangle$ to interact with her half of the EPR pair. This step is necessary. Indeed, if Alice simply performed a measurement in the computational basis, the quantum state $|\psi\rangle$ would collapse onto $|0\rangle$ or $|1\rangle$ and Alice would not obtain enough information to reconstruct it. The way out is a measurement in the Bell basis, given by the states $|\phi^+\rangle$, $|\phi^-\rangle$, $|\psi^+\rangle$ and $|\psi^-\rangle$ defined in subsection 3.4.1. These states constitute a complete orthonormal set and therefore one can expand the states of the computational basis over this basis. This gives

$$
\begin{cases}
|00\rangle = \tfrac{1}{\sqrt{2}} \left(|\phi^+\rangle + |\phi^-\rangle \right), \\[4pt]
|11\rangle = \tfrac{1}{\sqrt{2}} \left(|\phi^+\rangle - |\phi^-\rangle \right), \\[4pt]
|01\rangle = \tfrac{1}{\sqrt{2}} \left(|\psi^+\rangle + |\psi^-\rangle \right), \\[4pt]
|10\rangle = \tfrac{1}{\sqrt{2}} \left(|\psi^+\rangle - |\psi^-\rangle \right).
\end{cases} \qquad (4.36)
$$

We insert these relations into Eq. (4.35), obtaining

$$
\begin{aligned}
|\psi\rangle \otimes |\psi^+\rangle &= \tfrac{\alpha}{2}\big(|\phi^+\rangle + |\phi^-\rangle\big)|1\rangle + \tfrac{\alpha}{2}\big(|\psi^+\rangle + |\psi^-\rangle\big)|0\rangle \\
&\quad + \tfrac{\beta}{2}\big(|\psi^+\rangle - |\psi^-\rangle\big)|1\rangle + \tfrac{\beta}{2}\big(|\phi^+\rangle - |\phi^-\rangle\big)|0\rangle \\
&= \tfrac{1}{2}|\psi^+\rangle\big(\alpha|0\rangle + \beta|1\rangle\big) + \tfrac{1}{2}|\psi^-\rangle\big(\alpha|0\rangle - \beta|1\rangle\big) \\
&\quad + \tfrac{1}{2}|\phi^+\rangle\big(\alpha|1\rangle + \beta|0\rangle\big) + \tfrac{1}{2}|\phi^-\rangle\big(\alpha|1\rangle - \beta|0\rangle\big). \quad (4.37)
\end{aligned}
$$

Therefore, Alice must perform a Bell measurement, obtaining one out of the four states $|\psi^\pm\rangle$ or $|\phi^\pm\rangle$, with equal probability $p = \tfrac{1}{4}$. Note that, as we have seen in Sec. 3.4, the Bell measurement can be transformed into a standard measurement in the computational basis, provided that one applies the unitary transformation

$$
(H \otimes I)\, \text{CNOT} \qquad\qquad (4.38)
$$

before the measurement. This transforms $|\phi^+\rangle$ to $|00\rangle$, $|\psi^+\rangle$ to $|01\rangle$, $|\phi^-\rangle$ to $|10\rangle$ and $|\psi^-\rangle$ to $|11\rangle$ (see again Sec. 3.4). Thus, Alice applies the unitary transformation (4.38) to the two qubits that she possesses. This leads to the following global state for the three qubits:

$$
\begin{aligned}
&\tfrac{1}{2}|01\rangle\big(\alpha|0\rangle + \beta|1\rangle\big) + \tfrac{1}{2}|11\rangle\big(\alpha|0\rangle - \beta|1\rangle\big) \\
&+ \tfrac{1}{2}|00\rangle\big(\alpha|1\rangle + \beta|0\rangle\big) + \tfrac{1}{2}|10\rangle\big(\alpha|1\rangle - \beta|0\rangle\big). \quad (4.39)
\end{aligned}
$$

3. Alice measures the two qubits in her possession in the computational basis. The four possible outcomes (00, 01, 10 and 11) give two bits of classical information. As can be seen from Eq. (4.37), if the outcome of Alice's measurement is 00, then the state of Bob's particle collapses onto $\alpha|1\rangle + \beta|0\rangle$. Analogously, outcomes 01, 10 and 11 leave the post-measurement state of Bob's particle in $\alpha|0\rangle + \beta|1\rangle$, $\alpha|1\rangle - \beta|0\rangle$ and $\alpha|0\rangle - \beta|1\rangle$, respectively.

4. Alice sends these two classical bits to Bob.

5. Bob receives these two bits of *classical information*, telling him which of the four possible outcomes of her measurement Alice obtained. Depending on this *classical* message, Bob performs one out of four possible unitary operations U on his qubit to recover the state $|\psi\rangle$. If Alice obtained 00, Bob performs $U = \sigma_x$. Analogously, 01, 10 and 11 drive $U = I$, $U = i\sigma_y$ and $U = \sigma_z$, respectively.

We stress that teleportation does not allow one to communicate quantum information faster than light. Indeed, Alice must send two bits of classical information to allow Bob to reconstruct the state $|\psi\rangle$. This information is transmitted by classical means, at a speed not greater than that of light. Note also that it is the information about the quantum state, the qubit, that passes from Alice to Bob and not the physical system itself. The physical systems implementing the qubit can be very different in Alice's and Bob's laboratories.

We also emphasize that teleportation is fully consistent with the no-cloning theorem. The quantum state $|\psi\rangle$ is in Bob's possession at the end of the teleportation process, but the original state is left in $|0\rangle$ or $|1\rangle$, depending on the result of Alice's measurement. The unknown quantum state $|\psi\rangle$ vanishes in one place and reappears in another.

It is also interesting to note that dense coding and quantum teleportation can be obtained by the same quantum circuit, "cut" in different positions (see Fig. 4.7).

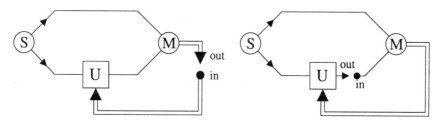

Fig. 4.7 Diagrams representing dense coding (left) and teleportation (right). Double lines represent two classical bits, single lines a quantum bit, S the EPR source, M the measurement process, U the unitary transformation driven by the two classical bits, "in" and "out" the input and output of the circuits.

Finally, we would like to emphasize that teleportation plays a very important role in a number of quantum computation protocols (Gottesman and Chuang, 1999; Knill *et al.*, 2001). It is a powerful tool for transfering quantum states from one system to another, as would be required in a quantum computer made of several independent units. In particular, it has been proved that teleportation, together with single-qubit operations, is sufficient to achieve universal quantum computation (Gottesman and Chuang, 1999).

Exercise 4.6 Study the quantum circuit of Fig. 4.8. It achieves teleportation via quantum computation (see Brassard *et al.*, 1998). Show that in the output the state $|\psi\rangle$ is recovered in the third line of the circuit. This circuit is sometimes called *intraportation*, since the CNOT gates are

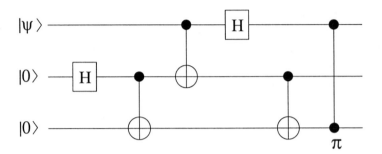

Fig. 4.8   A quantum circuit implementing intraportation. The last gate is a controlled phase shift through an angle π.

performed between the first-second and second-third qubits. Therefore, in order to implement these CNOT gates, the first two qubits cannot be arbitrarily far away from the third one.

Exercise 4.7   Study the circuit of Fig. 4.9 for the teleportation of an

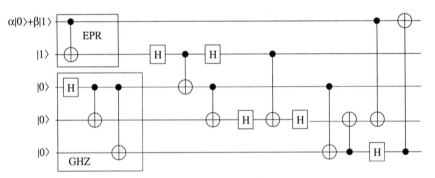

Fig. 4.9   A quantum circuit implementing the quantum teleportation of an entangled pair.

EPR pair (see Gorbachev and Trubilko, 2000). The first quantum gates generate the entangled state

$$\alpha \,|01\rangle + \beta \,|10\rangle \tag{4.40}$$

and the GHZ (Greenberger, Horne and Zeilinger) state

$$\tfrac{1}{\sqrt{2}} \left(|000\rangle + |111\rangle\right). \tag{4.41}$$

Show that at the end the EPR state (4.40) is recovered in the last two lines of the circuit in Fig. 4.9.

4.6   An overview of the experimental implementations

Several quantum optical experiments demonstrated the teleportation protocol, starting from the experiments performed with EPR pairs of photons in Rome and Innsbruck. Recently, a long distance implementation has been achieved in Geneva using a 2 km optical fibre. Besides these rapid developments in the quantum-optics arena, other realizations are particularly interesting from the viewpoint of quantum computation. Among them, NMR experiments have been successful in implementing teleportation. Moreover, there are proposals for teleporting atomic states and using quantum-dot systems for electron teleportation.

Most probably, quantum cryptography (or more precisely, quantum-key distribution) will be the first quantum information protocol to find commercial applications. Recent experimental advances in this field have been impressive and quantum cryptographic protocols have been demonstrated, using optical fibres, over distances of a few tens of kilometres at rates of the order of a thousand bits per second. Note that we refer to optical experiments, the physical realization of a qubit being a single photon. We should point out that these experiments use standard optical communication channels and that it is therefore not necessary to construct special-purpose cables.

The bottleneck in quantum communication via optical fibres is that the probability for both absorption losses and depolarization of photons grows exponentially with the length of the fibre. On the basis of present technology, it seems difficult to use optical fibres for quantum communication over distances larger than about 100 kilometres. To cover longer distances would require the use of quantum repeaters, that is, quantum purification schemes aimed at improving the fidelity of the transmitted photons (see Briegel *et al.*, 1998). Such quantum repeaters have not yet been demonstrated experimentally.

In a completely different approach, the qubits (photons) are transmitted through free-space. With this approach, quantum cryptography has been recently demonstrated over distances up to several kilometres. It is important to point out that the turbulence encountered along the optical path is comparable to the effective turbulence in earth-to-satellite transmission. Therefore, one can expect that in the near future it will become possible to use free-space photon transmission to distribute secret keys between parties located very far apart (say, in two different continents), using satellite-based links.

4.7   A guide to the bibliography

The no-cloning theorem is due to Dieks (1982) and to Wootters and Zurek (1982). This theorem does not forbid the existence of imperfect cloning machines, discussed, for instance, in Bužek and Hillery (1996), Gisin and Massar (1997) and Bruß *et al.* (1998).

A book on classical cryptography is Welsh (1997). The quantum cryptographic protocols discussed in Sec. 4.3 are due to Bennett and Brassard (1984) and Ekert (1991), see also Bennett *et al.* (1992). Another interesting protocol using non-orthogonal quantum states was introduced by Bennett (1992). A recent review of quantum cryptography is Gisin *et al.* (2002). Very readable introductions are Bennett *et al.* (1991), Bruß and Lütkenhaus (2000) and Lomonaco (2001). Quantum-key distribution has been demonstrates over fibre (for a review, see Gisin *et al.*, 2002) and free-space optical links (Buttler *et al.*, 2000; Kurtsiefer *et al.*, 2002). Moreover, it is possible to share entangled photon pairs over long distances, via fibres (Tapster *et al.*, 1994; Tittel *et al.*, 1998) or over a free-space path (Asplemeyer *et al.*, 2003).

Quantum teleportation was discovered by Bennett *et al.* (1993). The first experimental implementations were performed using quantum optics techniques (Boschi *et al.*, 1998; Bouwmeester *et al.*, 1997), continuous variables (Furusawa *et al.*, 1998) and NMR techniques (Nielsen *et al.*, 1998).

The quantum dense coding protocol is due to Bennett and Wiesner (1992) and an experimental implementation was performed by Mattle *et al.* (1996).

Appendix A

Solutions to the exercises

Chapter 1

Exercise 1.1

$$(a \uparrow b) \uparrow (a \uparrow b) = \overline{(a \uparrow b) \wedge (a \uparrow b)} = 1 - (a \uparrow b)^2$$

$$= 1 - (a \uparrow b) = 1 - (1 - ab) = ab = a \wedge b. \quad \text{(A.1a)}$$

$$(a \uparrow a) \uparrow (b \uparrow b) = (1 - a \wedge a) \uparrow (1 - b \wedge b) = (1 - a^2) \uparrow (1 - b^2)$$

$$= (1 - a) \uparrow (1 - b) = \overline{(1 - a) \wedge (1 - b)}$$

$$= 1 - (1 - a)(1 - b) = a + b - ab = a \vee b. \quad \text{(A.1b)}$$

Exercise 1.2 If we set as input $a = b = 1$, then we obtain as output $c' = \bar{c}$. If we set $c = 0$, then $c' = a \wedge b$. We can then obtain the OR gate as follows. From the NOT gate we obtain $\bar{a}$ and $\bar{b}$; we then perform a Toffoli gate on inputs $\bar{a}$, $\bar{b}$ and $c = 1$, giving $c' = a \vee b$.

Exercise 1.3 It is sufficient to show that from the Fredkin gate we can construct the universal set of gates AND, OR, NOT and FANOUT. If we set $b = 0$ and $c = 1$ as input of the Fredkin gate, then $b' = a$ and $c' = \bar{a}$. Therefore, we obtain simultaneously the FANOUT and NOT gates. Setting $c = 0$, we obtain $c' = a \wedge b$. The OR gate can be constructed from the AND and NOT gates by means of a De Morgan identity: $a \vee b = \overline{\bar{a} \wedge \bar{b}}$.

Chapter 2

Exercise 2.1 Assume that $\{|\alpha_{i1}\rangle, |\alpha_{i2}\rangle, \ldots, |\alpha_{ik}\rangle\}$ are eigenvectors of the linear operator A corresponding to the eigenvalue α_i and that $|\alpha_j\rangle$ is an eigenvector corresponding to the eigenvalue α_j. Furthermore, assume that $\alpha_i \neq \alpha_j$ and that the eigenvectors $\{|\alpha_{il}\rangle\}$ and $|\alpha_j\rangle$ are linearly dependent; that is,

$$|\lambda_j\rangle = \sum_{l=1}^{k} c_l |\alpha_{il}\rangle. \tag{A.2}$$

We then have

$$A|\alpha_j\rangle = \sum_{l=1}^{k} c_l A|\alpha_{il}\rangle = \sum_{l=1}^{k} c_l \alpha_i |\alpha_{il}\rangle = \alpha_i |\alpha_j\rangle. \tag{A.3}$$

Since $A|\alpha_j\rangle = \alpha_j|\alpha_j\rangle$, we find $\alpha_i = \alpha_j$, which contradicts the hypothesis that the eigenvalues α_i and α_j are distinct.

Exercise 2.2   The linearity of the operators A and B implies that

$$[(A+B)^\dagger]_{ij} = (A+B)^*_{ji} = A^*_{ji} + B^*_{ji} = A^\dagger_{ij} + B^\dagger_{ij}. \tag{A.4}$$

Since

$$(AB)_{ij} = \sum_k A_{ik} B_{kj}, \tag{A.5}$$

we have

$$(AB)^\dagger_{ij} = (AB)^*_{ji} = \sum_k A^*_{jk} B^*_{ki}. \tag{A.6}$$

Furthermore, we obtain

$$(B^\dagger A^\dagger)_{ij} = \sum_k (B^\dagger)_{ik} (A^\dagger)_{kj} = \sum_k B^*_{ki} A^*_{jk} = \sum_k A^*_{jk} B^*_{ki}, \tag{A.7}$$

and therefore comparison of Eqs. (A.6) and (A.7) proves that $(AB)^\dagger = B^\dagger A^\dagger$. Finally, we obtain

$$[(A^\dagger)^\dagger]_{ij} = (A^\dagger)^*_{ji} = [A^*_{ij}]^* = A_{ij}. \tag{A.8}$$

Exercise 2.3   From Eq. (2.55) we find that the projector P has matrix elements

$$P_{ij} = \langle i|P|j\rangle = \sum_l \langle i|\alpha_l\rangle\langle \alpha_l|j\rangle, \tag{A.9}$$

where $\{|i\rangle\}$ is a basis for the Hilbert space. Therefore,

$$P^*_{ji} = \sum_l \langle j|\alpha_l\rangle^* \langle \alpha_l|i\rangle^* = \sum_l \langle \alpha_l|j\rangle\langle i|\alpha_l\rangle = P_{ij}. \tag{A.10}$$

Hence, a projector is an Hermitian operator. If we exclude the trivial case in which $P = I$, a projector cannot be inverted, since it has zero eigenvalues, corresponding to the eigenvectors residing in the subspace orthogonal to the subspace onto which P projects. Therefore, $\det P = 0$ and P cannot be inverted.

Exercise 2.4   We can immediately check that the Pauli matrices are Hermitian: since $\sigma_i = (\sigma_i^T)^\star$ $(i = x, y, z)$. By computing σ_i^{-1}, we can also check that $\sigma_i^{-1} = (\sigma_i^T)^\star$ and therefore the Pauli matrices are also unitary.

Exercise 2.5   Let us call $|\phi\rangle$ the vector obtained by application of the operator A to a generic vector $|\psi\rangle$; that is, $|\phi\rangle = A|\psi\rangle$. All vectors are transformed by means of the matrix S^{-1} (see Eq. (2.76)), in particular $|\psi'\rangle = S^{-1}|\psi\rangle$ and $|\phi'\rangle = S^{-1}|\phi\rangle$. Therefore,

$$S\,|\phi\rangle \,=\, SA\,|\psi\rangle \,=\, SAS^{-1}S\,|\psi\rangle\,, \tag{A.11}$$

where we have used the relation $S^{-1}S = I$. Finally, we obtain

$$|\phi'\rangle \,=\, SAS^{-1}\,|\psi'\rangle \,=\, A'\,|\psi'\rangle\,, \tag{A.12}$$

which implies the final result

$$A' \,=\, S^{-1}AS\,. \tag{A.13}$$

Exercise 2.6   We have $\sigma_i' = S^{-1}\sigma_i S$ $(i = x, y, z)$, where the matrix S relating the old basis vectors $\{|0\rangle, |1\rangle\}$ to the new $\{|+\rangle, |-\rangle\}$ is given by Eq. (2.87). Note that the matrix S is self-inverse, that is, $S^{-1} = S$. Therefore, we obtain

$$\sigma_x' \,=\, \frac{1}{\sqrt{2}}\begin{bmatrix} 1 & 1 \\ 1 & -1 \end{bmatrix}\begin{bmatrix} 0 & 1 \\ 1 & 0 \end{bmatrix}\frac{1}{\sqrt{2}}\begin{bmatrix} 1 & 1 \\ 1 & -1 \end{bmatrix} \,=\, \begin{bmatrix} 1 & 0 \\ 0 & -1 \end{bmatrix} \,=\, \sigma_z\,. \tag{A.14}$$

Similarly, we obtain $\sigma_y' = -\sigma_y$ and $\sigma_z' = \sigma_x$.

Exercise 2.7   Since

$$A^\dagger = A\,, \qquad B^\dagger = B\,, \tag{A.15}$$

we obtain:

$$\{i[A, B]\}^\dagger \,=\, \{iAB - iBA\}^\dagger \,=\, -iB^\dagger A^\dagger + iA^\dagger B^\dagger \,=\, i[A, B]\,. \tag{A.16}$$

Exercise 2.8

$$\sigma_x \otimes \sigma_y \,=\, \begin{bmatrix} 0 & 1 \\ 1 & 0 \end{bmatrix} \otimes \begin{bmatrix} 0 & -i \\ i & 0 \end{bmatrix} \,=\, \begin{bmatrix} 0 & 0 & 0 & -i \\ 0 & 0 & i & 0 \\ 0 & -i & 0 & 0 \\ i & 0 & 0 & 0 \end{bmatrix}. \tag{A.17}$$

$$I \otimes \sigma_x \,=\, \begin{bmatrix} 1 & 0 \\ 0 & 1 \end{bmatrix} \otimes \begin{bmatrix} 0 & 1 \\ 1 & 0 \end{bmatrix} \,=\, \begin{bmatrix} 0 & 1 & 0 & 0 \\ 1 & 0 & 0 & 0 \\ 0 & 0 & 0 & 1 \\ 0 & 0 & 1 & 0 \end{bmatrix}. \tag{A.18}$$

Exercise 2.9   Any unitary operator U is normal and therefore diagonalizable. Hence, we can write its spectral decomposition as

$$U = \sum_j \lambda_j |j\rangle\langle j| \,. \tag{A.19}$$

Since unitary operators preserve the inner product between vectors we have

$$\langle j|U^\dagger U|j\rangle = \lambda_j \langle j|j\rangle \,, \tag{A.20}$$

which implies $|\lambda_j| = 1$. Thus, Eq. (A.19) can be rewritten as follows:

$$U = \sum_j e^{i\alpha_j} |j\rangle\langle j|, \tag{A.21}$$

where α_j is a real number. We may now define the operator

$$A = \sum_j \alpha_j |j\rangle\langle j|. \tag{A.22}$$

Since A is already written in its diagonal basis, the operator e^{iA} is simply given by

$$e^{iA} = \sum_j e^{i\alpha_j} |j\rangle\langle j|, \tag{A.23}$$

and therefore $e^{iA} = U$. Let us prove that A is Hermitian. We have

$$U = \sum_{n=0}^{\infty} \frac{(iA)^n}{n!} \,, \tag{A.24}$$

and therefore

$$U^\dagger = \sum_{n=0}^{\infty} \left[\frac{(iA)^n}{n!} \right]^\dagger = \sum_{n=0}^{\infty} \frac{(-iA^\dagger)^n}{n!} = e^{-iA^\dagger}. \tag{A.25}$$

Since $U^{-1} = e^{-iA}$, the condition $U^\dagger = U^{-1}$ is satisfied when $A^\dagger = A$, namely, when A is an Hermitian operator.

Exercise 2.10   The Heisenberg uncertainty principle tells us that

$$\Delta\sigma_x \, \Delta\sigma_y \geq \tfrac{1}{2} |\langle 0|[\sigma_x, \sigma_y]|0\rangle|. \tag{A.26}$$

After evaluation of the commutator $[\sigma_x, \sigma_y]$, we obtain

$$\Delta\sigma_x \, \Delta\sigma_y \geq \tfrac{1}{2} \left| \begin{bmatrix} 1 & 0 \end{bmatrix} \begin{bmatrix} 2i & 0 \\ 0 & -2i \end{bmatrix} \begin{bmatrix} 1 \\ 0 \end{bmatrix} \right| = 1. \tag{A.27}$$

Exercise 2.11 The first apparatus prepares the state $|+\rangle_z = |1\rangle$. The state $|1\rangle$ then enters the second apparatus, which prepares the state $|+\rangle_y = \frac{1}{\sqrt{2}}(-i|0\rangle + |1\rangle)$. Finally, the third apparatus analyzes the state $|+\rangle_y$: it measures σ_z and obtains the two possible outcomes $|+\rangle_z = |0\rangle$ or $|-\rangle_z = |1\rangle$ with equal probabilities p_0 and p_1, since

$$p_0 = \left| \langle 0|+\rangle_y \right|^2 = \tfrac{1}{2}, \qquad p_1 = \left| \langle 1|+\rangle_y \right|^2 = \tfrac{1}{2}. \tag{A.28}$$

Exercise 2.12 The solution to the Schrödinger equation (2.164) is given by

$$|\psi(t)\rangle = \begin{bmatrix} a(t) \\ b(t) \end{bmatrix} = U(t) \begin{bmatrix} a(0) \\ b(0) \end{bmatrix}, \tag{A.29}$$

where the unitary time-evolution operator is

$$U(t) = \exp\left[-\frac{i}{\hbar} H t \right], \tag{A.30}$$

and

$$H = -\mu \boldsymbol{H} \cdot \boldsymbol{\sigma} \tag{A.31}$$

is the Hamiltonian of the system. Let us compute explicitly $U(t)$. We have

$$-\frac{i}{\hbar} H t = \frac{i\mu t}{\hbar} (\boldsymbol{H} \cdot \boldsymbol{\sigma}) = i\alpha (\boldsymbol{n} \cdot \boldsymbol{\sigma}), \tag{A.32}$$

where we have defined

$$\alpha = \frac{\mu t}{\hbar} \sqrt{H_x^2 + H_y^2 + H_z^2}, \qquad \boldsymbol{n} = \frac{1}{\sqrt{H_x^2 + H_y^2 + H_z^2}} (H_x, H_y, H_z). \tag{A.33}$$

Performing a Taylor expansion of the operator $U(t)$ we obtain

$$\begin{aligned} U(t) &= e^{i\alpha\, \boldsymbol{n} \cdot \boldsymbol{\sigma}} \\ &= \left[I - \tfrac{1}{2!} \alpha^2 (\boldsymbol{n} \cdot \boldsymbol{\sigma})^2 + \dots \right] + i \left[\alpha - \tfrac{1}{3!} \alpha^3 (\boldsymbol{n} \cdot \boldsymbol{\sigma})^3 + \dots \right] \\ &= \cos\alpha\, I + i \sin\alpha\, (\boldsymbol{n} \cdot \boldsymbol{\sigma}), \end{aligned} \tag{A.34}$$

where the last equality follows since $(\boldsymbol{n} \cdot \boldsymbol{\sigma})^2 = I$. Hence, we have

$$U(t) = \begin{bmatrix} \cos\alpha + i\sin\alpha\, n_z & \sin\alpha\,(n_y + in_x) \\ \sin\alpha\,(-n_y + in_x) & \cos\alpha - i\sin\alpha\, n_z \end{bmatrix}, \tag{A.35}$$

where n_x, n_y, and n_z are the Cartesian components of the unit vector $\boldsymbol{n}$. The average values of the Pauli operators are computed as follows:

$$\langle \sigma_i \rangle = \langle \psi(t)| \sigma_i |\psi(t)\rangle = \langle \psi(0)| U^\dagger \sigma_i U |\psi(0)\rangle. \tag{A.36}$$

To obtain the state $|1\rangle$ starting from the initial state $|0\rangle$, we can choose a magnetic field directed along the x axis. This means

$$\boldsymbol{H} = (H_x, 0, 0) \qquad \text{and} \qquad \boldsymbol{n} = (1, 0, 0). \tag{A.37}$$

We require that at time $\bar{t}$ the wave vector $|\psi(\bar{t})\rangle = U(\bar{t})|0\rangle$ coincide with $|1\rangle$; that is,

$$\begin{bmatrix} 0 \\ 1 \end{bmatrix} = U \begin{bmatrix} 1 \\ 0 \end{bmatrix} = \begin{bmatrix} \cos(\alpha(\bar{t})) & i\sin(\alpha(\bar{t})) \\ i\sin(\alpha(\bar{t})) & \cos(\alpha(\bar{t})) \end{bmatrix} \begin{bmatrix} 1 \\ 0 \end{bmatrix}. \tag{A.38}$$

This condition is fulfilled (up to an overall phase factor of no physical significance) when $\cos(\alpha(\bar{t})) = 0$, which is first satisfied after a time $\bar{t}$ such that

$$\alpha(\bar{t}) = \frac{\mu |H_x| \bar{t}}{\hbar} = \frac{\pi}{2}. \tag{A.39}$$

Exercise 2.13   A two-qubit state $|\psi\rangle$ is separable if and only if we can write it as follows:

$$|\psi\rangle = \left(\alpha|0\rangle + \beta|1\rangle\right) \otimes \left(\gamma|0\rangle + \delta|1\rangle\right), \tag{A.40}$$

with α, β, γ and δ complex coefficients satisfying the normalization conditions $|\alpha|^2 + |\beta|^2 = 1$ and $|\gamma|^2 + |\delta|^2 = 1$. If the state $|\psi\rangle$ is given by (2.171), then the separability condition (A.40) implies $\alpha\gamma = \frac{1}{\sqrt{2}}$, $\beta\delta = \frac{1}{\sqrt{2}}$, $\alpha\delta = 0$ and $\beta\gamma = 0$. As these four relations cannot be satisfied simultaneously, the state must be entangled.

Exercise 2.14   After insertion of the explicit expressions for $|+\rangle_u$ and $|-\rangle_u$, given by Eq. (2.160), we obtain:

$$|+\rangle_u|-\rangle_u - |-\rangle_u|+\rangle_u$$

$$= \frac{1}{\sqrt{2}} \begin{bmatrix} \cos\frac{\theta}{2} e^{-i\phi/2} \\ \sin\frac{\theta}{2} e^{i\phi/2} \end{bmatrix} \otimes \frac{1}{\sqrt{2}} \begin{bmatrix} -\sin\frac{\theta}{2} e^{-i\phi/2} \\ \cos\frac{\theta}{2} e^{i\phi/2} \end{bmatrix}$$

$$- \frac{1}{\sqrt{2}} \begin{bmatrix} -\sin\frac{\theta}{2} e^{-i\phi/2} \\ \cos\frac{\theta}{2} e^{i\phi/2} \end{bmatrix} \otimes \frac{1}{\sqrt{2}} \begin{bmatrix} \cos\frac{\theta}{2} e^{-i\phi/2} \\ \sin\frac{\theta}{2} e^{i\phi/2} \end{bmatrix} = \frac{1}{\sqrt{2}} \begin{bmatrix} 0 \\ 1 \\ -1 \\ 0 \end{bmatrix}. \tag{A.41}$$

The final state is indeed equal to the singlet state $\frac{1}{\sqrt{2}}\left(|01\rangle - |10\rangle\right)$, independently of the direction of $\boldsymbol{u}$.

Exercise 2.15   (a) Measurement of σ_z for the first particle. It is convenient to rewrite the state (2.178) as

$$|\psi\rangle = \sqrt{|\alpha|^2 + |\beta|^2} \, |0\rangle \otimes \frac{\alpha|0\rangle + \beta|1\rangle}{\sqrt{|\alpha|^2 + |\beta|^2}} + \sqrt{|\gamma|^2 + |\delta|^2} \, |1\rangle \otimes \frac{\gamma|0\rangle + \delta|1\rangle}{\sqrt{|\gamma|^2 + |\delta|^2}}. \tag{A.42}$$

Therefore, a measurement of σ_z for the first particle results in outcome $+1$ with probability $(|\alpha|^2 + |\beta|^2)$ and outcome -1 with probability $(|\gamma|^2 + |\delta|^2)$. Furthermore, after a measurement with result $\sigma_z = +1$, the state of the second particle collapses onto the state

$$|\phi^{(0)}\rangle_2 = \frac{\alpha|0\rangle + \beta|1\rangle}{\sqrt{|\alpha|^2 + |\beta|^2}} \; ; \tag{A.43a}$$

if instead the result is $\sigma_z = -1$, the state of the second particle collapses onto

$$|\phi^{(1)}\rangle_2 = \frac{\gamma|0\rangle + \delta|1\rangle}{\sqrt{|\gamma|^2 + |\delta|^2}} \; . \tag{A.43b}$$

(b) Measurement of σ_x for the first particle. Taking into account that

$$|0\rangle = \tfrac{1}{\sqrt{2}}\left(|+\rangle + |-\rangle\right), \qquad |1\rangle = \tfrac{1}{\sqrt{2}}\left(|+\rangle - |-\rangle\right), \tag{A.44}$$

we can rewrite the state (2.178) as

$$
\begin{aligned}
|\psi\rangle &= \tfrac{1}{\sqrt{2}}\left(|+\rangle + |-\rangle\right)\left(\alpha|0\rangle + \beta|1\rangle\right) + \tfrac{1}{\sqrt{2}}\left(|+\rangle - |-\rangle\right)\left(\gamma|0\rangle + \delta|1\rangle\right) \\
&= \sqrt{\frac{|\alpha + \gamma|^2 + |\beta + \delta|^2}{2}}\,|+\rangle \otimes \frac{(\alpha + \gamma)\,|0\rangle + (\beta + \delta)\,|1\rangle}{\sqrt{|\alpha + \gamma|^2 + |\beta + \delta|^2}} \\
&\quad + \sqrt{\frac{|\alpha - \gamma|^2 + |\beta - \delta|^2}{2}}\,|-\rangle \otimes \frac{(\alpha - \gamma)\,|0\rangle + (\beta - \delta)\,|1\rangle}{\sqrt{|\alpha - \gamma|^2 + |\beta - \delta|^2}} \; .
\end{aligned}
\tag{A.45}
$$

Then, analogously to the previous case, we can compute the probability of obtaining $\sigma_x = +1$ or $\sigma_x = -1$ and the corresponding wave vectors onto which the state of the second particle collapses after the measurement.

Exercise 2.16   It is easy to check that

$$
\begin{aligned}
\langle 0|\,\boldsymbol{\sigma}\cdot\boldsymbol{r}\,|0\rangle &= z, & \langle 1|\,\boldsymbol{\sigma}\cdot\boldsymbol{r}\,|1\rangle &= -z, \\
\langle 0|\,\boldsymbol{\sigma}\cdot\boldsymbol{r}\,|1\rangle &= x - iy, & \langle 1|\,\boldsymbol{\sigma}\cdot\boldsymbol{r}\,|0\rangle &= x + iy,
\end{aligned}
\tag{A.46}
$$

where $\boldsymbol{\sigma} = (\sigma_x, \sigma_y, \sigma_z)$ and $\boldsymbol{r} = (x, y, z)$. Using equalities (A.46), we obtain

$$\langle\psi|(\boldsymbol{\sigma}^{(A)} \cdot \boldsymbol{a})(\boldsymbol{\sigma}^{(B)} \cdot \boldsymbol{b})|\psi\rangle$$

$$= \tfrac{1}{2}\left(\langle 01| - \langle 10|\right)\left(\boldsymbol{\sigma}^{(A)} \cdot \boldsymbol{a}\right)\left(\boldsymbol{\sigma}^{(B)} \cdot \boldsymbol{b}\right)\left(|01\rangle - |10\rangle\right)$$

$$= \tfrac{1}{2}\langle 0|\,\boldsymbol{\sigma}^{(A)} \cdot \boldsymbol{a}\,|0\rangle\langle 1|\,\boldsymbol{\sigma}^{(B)} \cdot \boldsymbol{b}\,|1\rangle - \tfrac{1}{2}\langle 0|\,\boldsymbol{\sigma}^{(A)} \cdot \boldsymbol{a}\,|1\rangle\langle 1|\,\boldsymbol{\sigma}^{(B)} \cdot \boldsymbol{b}\,|0\rangle$$

$$\quad - \tfrac{1}{2}\langle 1|\,\boldsymbol{\sigma}^{(A)} \cdot \boldsymbol{a}\,|0\rangle\langle 0|\,\boldsymbol{\sigma}^{(B)} \cdot \boldsymbol{b}\,|1\rangle + \tfrac{1}{2}\langle 1|\,\boldsymbol{\sigma}^{(A)} \cdot \boldsymbol{a}\,|1\rangle\langle 0|\,\boldsymbol{\sigma}^{(B)} \cdot \boldsymbol{b}\,|0\rangle$$

$$= -\boldsymbol{a} \cdot \boldsymbol{b}, \tag{A.47}$$

where $|\psi\rangle$ is the singlet state (2.175).

Exercise 2.17   We can always choose the arbitrary phases multiplying the basis states for the two spin-$\tfrac{1}{2}$ particles so that α and β are real and positive. Let us compute the correlator

$$C(\boldsymbol{a}, \boldsymbol{b}) = \langle\psi|(\boldsymbol{\sigma}^{(A)} \cdot \boldsymbol{a})(\boldsymbol{\sigma}^{(B)} \cdot \boldsymbol{b})|\psi\rangle$$

$$= \left(\alpha\langle 00| + \beta\langle 11|\right)\left(\boldsymbol{\sigma}^{(A)} \cdot \boldsymbol{a}\right)\left(\boldsymbol{\sigma}^{(B)} \cdot \boldsymbol{b}\right)\left(\alpha|00\rangle + \beta|11\rangle\right)$$

$$= \alpha^2 \langle 0|\,\boldsymbol{\sigma}^{(A)} \cdot \boldsymbol{a}\,|0\rangle\langle 0|\,\boldsymbol{\sigma}^{(B)} \cdot \boldsymbol{b}\,|0\rangle + \beta^2 \langle 1|\,\boldsymbol{\sigma}^{(A)} \cdot \boldsymbol{a}\,|1\rangle\langle 1|\,\boldsymbol{\sigma}^{(B)} \cdot \boldsymbol{b}\,|1\rangle$$

$$\quad + 2\alpha\beta\, \mathrm{Re}\left(\langle 0|\,\boldsymbol{\sigma}^{(A)} \cdot \boldsymbol{a}\,|1\rangle\langle 0|\,\boldsymbol{\sigma}^{(B)} \cdot \boldsymbol{b}\,|1\rangle\right)$$

$$= z_a z_b + 2\alpha\beta(x_a x_b - y_a y_b), \tag{A.48}$$

where $\boldsymbol{a} = (x_a, y_a, z_a)$ and $\boldsymbol{b} = (x_b, y_b, z_b)$. If we consider the set of directions $\boldsymbol{a} = (1, 0, 0)$, $\boldsymbol{a}' = (0, 0, 1)$, $\boldsymbol{b} = (x_b, 0, z_b)$ and $\boldsymbol{b}' = (-x_b, 0, z_b)$, we obtain

$$\left|C(\boldsymbol{a}, \boldsymbol{b}) - C(\boldsymbol{a}', \boldsymbol{b}')\right| + \left|C(\boldsymbol{a}', \boldsymbol{b}) + C(\boldsymbol{a}', \boldsymbol{b}')\right| = 2(z_b + 2\alpha\beta x_b). \tag{A.49}$$

Therefore, the CHSH inequality is violated if $z_b + 2\alpha\beta x_b > 1$. If θ_b denotes the angle between the unit vector $\boldsymbol{b}$ and the z axis, we obtain

$$z_b + 2\alpha\beta x_b = \cos\theta_b + 2\alpha\beta\sin\theta_b = 1 + 2\alpha\beta\theta_b + O(\theta^2), \tag{A.50}$$

which, insofar as α and β are both different from zero, is larger than 1, provided that θ_b is positive and small enough. Therefore, the violation of Bell's inequalities is a generic feature of entangled states.

Chapter 3

Exercise 3.1 It is convenient to perform a rigid rotation of the two states, $|\psi_1\rangle \to |\psi_1'\rangle = R|\psi_1\rangle$ and $|\psi_2\rangle \to |\psi_2'\rangle = R|\psi_2\rangle$, with R a rotation matrix, in such a way that one of the rotated states, for example $|\psi_1'\rangle$, coincides with the north pole of the Bloch sphere. Then we have $|\psi_1'\rangle = |0\rangle$ and $|\psi_2'\rangle = \cos\frac{\theta}{2}|0\rangle + e^{i\phi}\sin\frac{\theta}{2}|1\rangle$, where the angle θ and ϕ parametrize the position of the state vector $|\psi_2'\rangle$ on the Bloch sphere (see Eq. (3.4)). We note that θ is also the angle between the Bloch vectors $|\psi_1\rangle$ and $|\psi_2\rangle$. We have

$$F = |\langle\psi_1|\psi_2\rangle|^2 = |\langle\psi_1'|\psi_2'\rangle|^2 = \left| \begin{bmatrix} 1 & 0 \end{bmatrix} \begin{bmatrix} \cos\frac{\theta}{2} \\ e^{i\phi}\sin\frac{\theta}{2} \end{bmatrix} \right|^2 = \cos^2\frac{\theta}{2}. \qquad (A.51)$$

Exercise 3.2 We start by noting that

$$H\,R_z(\alpha)\,H = \begin{bmatrix} \cos\frac{\alpha}{2} & i\sin\frac{\alpha}{2} \\ i\sin\frac{\alpha}{2} & \cos\frac{\alpha}{2} \end{bmatrix} = \cos\frac{\alpha}{2}\,I + i\sin\frac{\alpha}{2}\,\sigma_x \qquad (A.52)$$

performs a rotation through an angle $-\alpha$ about the x-axis. The chain of rotations written in Eq. (3.27) act as follows:

(i) $R_z(-\frac{\pi}{2}-\phi_1)$ is a rotation of the Bloch sphere through an angle $-\frac{\pi}{2}-\phi_1$ about the z axis. Thus, the state parametrized on the Bloch sphere by (θ_1,ϕ_1) is moved into the state $(\theta_1,-\frac{\pi}{2})$. This state belongs to the plane $(-y,z)$.

(ii) As shown in Eq. (A.52), $H\,R_z(\theta_2-\theta_1)\,H$ rotates the Bloch sphere through an angle $\theta_1-\theta_2$ about the x axis. Thus, the state vector becomes $(\theta_2,-\frac{\pi}{2})$.

(iii) $R_z(\frac{\pi}{2}+\phi_2)$ is a rotation through an angle $\frac{\pi}{2}+\phi_2$ about the z axis and therefore leads to the final (θ_2,ϕ_2).

Exercise 3.3 After computation of $|\psi'\rangle = R_x(\delta)|\psi\rangle$, we obtain

$$\begin{cases} x' = x\,, \\ y' = y\cos\delta - z\sin\delta\,, \\ z' = y\sin\delta + z\cos\delta\,, \end{cases} \qquad (A.53)$$

where (x,y,z) and (x',y',z') denote the Cartesian coordinates of the vectors $|\psi\rangle$ and $|\psi'\rangle$. It is evident that (A.53) represents a counterclockwise rotation through an angle δ about the x axis. Likewise we can compute $|\psi'\rangle = R_y(\delta)|\psi\rangle$, obtaining

$$\begin{cases} x' = x\cos\delta + z\sin\delta\,, \\ y' = y\,, \\ z' = -x\sin\delta + z\cos\delta\,, \end{cases} \qquad (A.54)$$

which represent a counterclockwise rotation through an angle δ about the y axis.

Exercise 3.4   A simple comparison between the matrix U_1, Eq. (3.12), and the rotation matrix $R_y(\delta)$, Eq. (3.33), shows that $U_1 = R_y(-\frac{\pi}{2})$. Similarly, a comparison of U_2, Eq. (3.14), and $R_x(\delta)$, Eq. (3.32), shows that $U_2 = R_x(\frac{\pi}{2})$.

Exercise 3.5   A generic 2×2 unitary matrix U can be seen as a rotation through an angle δ about some axis of the Bloch sphere (this axis is directed along the unit vector $\boldsymbol{n} = (n_x, n_y, n_z)$). Therefore, we can express U according to Eq. (3.38). It follows that

$$\sqrt{U} = \cos\left(\tfrac{\delta}{4}\right) I - i \sin\left(\tfrac{\delta}{4}\right) (\boldsymbol{n} \cdot \boldsymbol{\sigma}). \tag{A.55}$$

Exercise 3.6   We get by direct computation

$$
\begin{aligned}
(\boldsymbol{a} \cdot \boldsymbol{\sigma})(\boldsymbol{b} \cdot \boldsymbol{\sigma}) &= (a_x \sigma_x + a_y \sigma_y + a_z \sigma_z)(b_x \sigma_x + b_y \sigma_y + b_z \sigma_z) \\
&= a_x b_x\, \sigma_x^2 + a_y b_y\, \sigma_y^2 + a_z b_z\, \sigma_z^2 + (a_x b_y - a_y b_x)\sigma_x \sigma_y \\
&\quad + (a_y b_z - a_z b_y)\sigma_y \sigma_z + (a_z b_x - a_x b_z)\sigma_z \sigma_x \\
&= a_x b_x\, I + a_y b_y\, I + a_z b_z\, I + (a_x b_y - a_y b_x) i \sigma_z \\
&\quad + (a_y b_z - a_z b_y) i \sigma_x + (a_z b_x - a_x b_z) i \sigma_y \\
&= (\boldsymbol{a} \cdot \boldsymbol{b}) I + i \boldsymbol{\sigma} \cdot (\boldsymbol{a} \times \boldsymbol{b}). \tag{A.56}
\end{aligned}
$$

Note that we have taken advantage of the following properties of the Pauli matrices: (i) $\sigma_y \sigma_x = -\sigma_y \sigma_x$, $\sigma_z \sigma_y = -\sigma_y \sigma_z$ and $\sigma_x \sigma_z = -\sigma_z \sigma_x$ (that is, the Pauli matrices anti-commute); (ii) $\sigma_x^2 = \sigma_y^2 = \sigma_z^2 = I$; and (iii) $\sigma_x \sigma_y = i \sigma_z$, $\sigma_y \sigma_z = i \sigma_x$ and $\sigma_z \sigma_x = i \sigma_y$.

Exercise 3.7   The state vector (3.45) can be equivalently rewritten as

$$|\psi\rangle = a \left\{ |00\rangle + b_0\, e^{i\phi_0}|01\rangle + b_1\, e^{i\phi_1}|10\rangle + b_1 b_0\, e^{i(\phi_1 + \phi_0)}|11\rangle \right\}. \tag{A.57}$$

The application of the CNOT gate to this state leads to

$$\text{CNOT}\,|\psi\rangle = a \left\{ |00\rangle + b_0\, e^{i\phi_0}|01\rangle + b_1\, e^{i\phi_1}|11\rangle + b_1 b_0\, e^{i(\phi_1 + \phi_0)}|10\rangle \right\}, \tag{A.58}$$

which is separable if and only if $b_0\, e^{i\phi_0} = 1$. Thus, CNOT generates an entangled state if and only if at least one of the following two conditions is fulfilled:

$$b_0 \neq 1, \qquad \phi_0 \neq 0. \tag{A.59}$$

Exercise 3.8   The implementation of the generalized CNOT gate B (defined by Eq. (3.46)) by means of the standard CNOT and single-qubit (NOT) gates is shown in Fig. A.1. The first NOT gate flips the state of the control qubit and thus makes the standard CNOT gate act non-trivially only if the state of the control qubit was $|0\rangle$ at the beginning (before the NOT gate). The second NOT gate restores the original state of the target qubit. With the same procedure we

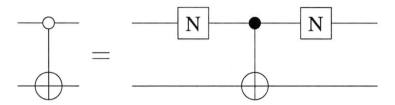

Fig. A.1 Decomposition of the generalized CNOT gate B into a standard CNOT gate and two NOT gates, denoted by N.

obtain the generalized CNOT gate D from C. Finally, we note that the NOT gates are implemented by the Pauli matrix σ_x.

In order to prove the equality between the two circuits shown in Fig. 3.3, we must verify that

$$C = H^{\otimes 2} A H^{\otimes 2}, \tag{A.60}$$

where A is the standard CNOT gate, C a generalized CNOT and $H^{\otimes 2} \equiv H \otimes H$. We have

$$H^{\otimes 2} = \frac{1}{2} \begin{bmatrix} 1 & 1 & 1 & 1 \\ 1 & -1 & 1 & -1 \\ 1 & 1 & -1 & -1 \\ 1 & -1 & -1 & 1 \end{bmatrix}. \tag{A.61}$$

Then we can explicitly perform the matrix products in Eq. (A.60) and verify that

$$\begin{bmatrix} 1 & 0 & 0 & 0 \\ 0 & 1 & 0 & 0 \\ 0 & 0 & 0 & 1 \\ 0 & 0 & 1 & 0 \end{bmatrix} = H^{\otimes 2} \begin{bmatrix} 0 & 0 & 1 & 0 \\ 0 & 1 & 0 & 0 \\ 1 & 0 & 0 & 0 \\ 0 & 0 & 0 & 1 \end{bmatrix} H^{\otimes 2}. \tag{A.62}$$

Exercise 3.9 It is sufficient to compute matrix products. For instance, we can check that the SWAP gate is implemented by the circuit of Fig. 3.4. Indeed, we have

$$A C A = \text{SWAP}, \tag{A.63}$$

since

$$\begin{bmatrix} 1 & 0 & 0 & 0 \\ 0 & 1 & 0 & 0 \\ 0 & 0 & 0 & 1 \\ 0 & 0 & 1 & 0 \end{bmatrix} \begin{bmatrix} 1 & 0 & 0 & 0 \\ 0 & 0 & 0 & 1 \\ 0 & 0 & 1 & 0 \\ 0 & 1 & 0 & 0 \end{bmatrix} \begin{bmatrix} 1 & 0 & 0 & 0 \\ 0 & 1 & 0 & 0 \\ 0 & 0 & 0 & 1 \\ 0 & 0 & 1 & 0 \end{bmatrix} = \begin{bmatrix} 1 & 0 & 0 & 0 \\ 0 & 0 & 1 & 0 \\ 0 & 1 & 0 & 0 \\ 0 & 0 & 0 & 1 \end{bmatrix}. \tag{A.64}$$

Equivalently, we can apply the sequence of generalized CNOT gates (ACA) to the states of the computational basis and check that

$$A C A |00\rangle = |00\rangle = \text{SWAP} |00\rangle, \quad A C A |01\rangle = |10\rangle = \text{SWAP} |01\rangle,$$
$$A C A |10\rangle = |01\rangle = \text{SWAP} |10\rangle, \quad A C A |11\rangle = |11\rangle = \text{SWAP} |11\rangle. \tag{A.65}$$

Therefore, $ACA = \text{SWAP}$ since both operators are linear and have the same action on a set of basis vectors.

The 24 possible permutation matrices are given by

$$P_1 = \begin{bmatrix} 1 & 0 & 0 & 0 \\ 0 & 1 & 0 & 0 \\ 0 & 0 & 1 & 0 \\ 0 & 0 & 0 & 1 \end{bmatrix}, \quad P_2 = \begin{bmatrix} 1 & 0 & 0 & 0 \\ 0 & 1 & 0 & 0 \\ 0 & 0 & 0 & 1 \\ 0 & 0 & 1 & 0 \end{bmatrix}, \quad P_3 = \begin{bmatrix} 1 & 0 & 0 & 0 \\ 0 & 0 & 1 & 0 \\ 0 & 1 & 0 & 0 \\ 0 & 0 & 0 & 1 \end{bmatrix},$$

$$P_4 = \begin{bmatrix} 1 & 0 & 0 & 0 \\ 0 & 0 & 1 & 0 \\ 0 & 0 & 0 & 1 \\ 0 & 1 & 0 & 0 \end{bmatrix}, \quad P_5 = \begin{bmatrix} 1 & 0 & 0 & 0 \\ 0 & 0 & 0 & 1 \\ 0 & 1 & 0 & 0 \\ 0 & 0 & 1 & 0 \end{bmatrix}, \quad P_6 = \begin{bmatrix} 1 & 0 & 0 & 0 \\ 0 & 0 & 0 & 1 \\ 0 & 0 & 1 & 0 \\ 0 & 1 & 0 & 0 \end{bmatrix},$$

$$P_7 = \begin{bmatrix} 0 & 1 & 0 & 0 \\ 1 & 0 & 0 & 0 \\ 0 & 0 & 1 & 0 \\ 0 & 0 & 0 & 1 \end{bmatrix}, \quad P_8 = \begin{bmatrix} 0 & 1 & 0 & 0 \\ 1 & 0 & 0 & 0 \\ 0 & 0 & 0 & 1 \\ 0 & 0 & 1 & 0 \end{bmatrix}, \quad P_9 = \begin{bmatrix} 0 & 1 & 0 & 0 \\ 0 & 0 & 1 & 0 \\ 0 & 0 & 0 & 1 \\ 1 & 0 & 0 & 0 \end{bmatrix},$$

$$P_{10} = \begin{bmatrix} 0 & 1 & 0 & 0 \\ 0 & 0 & 1 & 0 \\ 1 & 0 & 0 & 0 \\ 0 & 0 & 0 & 1 \end{bmatrix}, \quad P_{11} = \begin{bmatrix} 0 & 1 & 0 & 0 \\ 0 & 0 & 0 & 1 \\ 0 & 0 & 1 & 0 \\ 1 & 0 & 0 & 0 \end{bmatrix}, \quad P_{12} = \begin{bmatrix} 0 & 1 & 0 & 0 \\ 0 & 0 & 0 & 1 \\ 1 & 0 & 0 & 0 \\ 0 & 0 & 1 & 0 \end{bmatrix},$$

$$P_{13} = \begin{bmatrix} 0 & 0 & 1 & 0 \\ 1 & 0 & 0 & 0 \\ 0 & 1 & 0 & 0 \\ 0 & 0 & 0 & 1 \end{bmatrix}, \quad P_{14} = \begin{bmatrix} 0 & 0 & 1 & 0 \\ 1 & 0 & 0 & 0 \\ 0 & 0 & 0 & 1 \\ 0 & 1 & 0 & 0 \end{bmatrix}, \quad P_{15} = \begin{bmatrix} 0 & 0 & 1 & 0 \\ 0 & 1 & 0 & 0 \\ 1 & 0 & 0 & 0 \\ 0 & 0 & 0 & 1 \end{bmatrix},$$

$$P_{16} = \begin{bmatrix} 0 & 0 & 1 & 0 \\ 0 & 1 & 0 & 0 \\ 0 & 0 & 0 & 1 \\ 1 & 0 & 0 & 0 \end{bmatrix}, \quad P_{17} = \begin{bmatrix} 0 & 0 & 1 & 0 \\ 0 & 0 & 0 & 1 \\ 1 & 0 & 0 & 0 \\ 0 & 1 & 0 & 0 \end{bmatrix}, \quad P_{18} = \begin{bmatrix} 0 & 0 & 1 & 0 \\ 0 & 0 & 0 & 1 \\ 0 & 1 & 0 & 0 \\ 1 & 0 & 0 & 0 \end{bmatrix},$$

$$P_{19} = \begin{bmatrix} 0 & 0 & 0 & 1 \\ 1 & 0 & 0 & 0 \\ 0 & 1 & 0 & 0 \\ 0 & 0 & 1 & 0 \end{bmatrix}, \quad P_{20} = \begin{bmatrix} 0 & 0 & 0 & 1 \\ 1 & 0 & 0 & 0 \\ 0 & 0 & 1 & 0 \\ 0 & 1 & 0 & 0 \end{bmatrix}, \quad P_{21} = \begin{bmatrix} 0 & 0 & 0 & 1 \\ 0 & 1 & 0 & 0 \\ 1 & 0 & 0 & 0 \\ 0 & 0 & 1 & 0 \end{bmatrix},$$

$$P_{22} = \begin{bmatrix} 0 & 0 & 0 & 1 \\ 0 & 1 & 0 & 0 \\ 0 & 0 & 1 & 0 \\ 1 & 0 & 0 & 0 \end{bmatrix}, \quad P_{23} = \begin{bmatrix} 0 & 0 & 0 & 1 \\ 0 & 0 & 1 & 0 \\ 1 & 0 & 0 & 0 \\ 0 & 1 & 0 & 0 \end{bmatrix}, \quad P_{24} = \begin{bmatrix} 0 & 0 & 0 & 1 \\ 0 & 0 & 1 & 0 \\ 0 & 1 & 0 & 0 \\ 1 & 0 & 0 & 0 \end{bmatrix}. \quad (A.66)$$

Note that $P_3 = \text{SWAP}$. These permutation matrices are implemented by means of generalized CNOT gates, as shown in Fig. A.2.

Exercise 3.10   It is easy to check by direct matrix multiplication that

$$\text{CMINUS} = (I \otimes H)\, \text{CNOT}\, (I \otimes H). \quad (A.67)$$

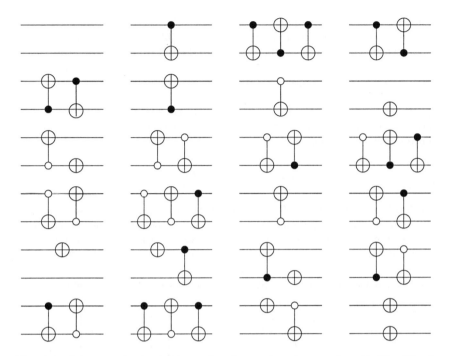

Fig. A.2 Quantum circuits implementing the permutation matrices (A.66). The $\oplus$ symbol, without a control qubit, denotes a simple NOT gate (see, *e.g.*, permutation P_8). Permutations are ordered from top to bottom and from left to right: the top line gives quantum circuits from P_1 (left) to P_4 (right), the second line (from the top) from P_5 (left) to P_8 (right) and so on.

Then the relation

$$\text{CNOT} = (I \otimes H)\,\text{CMINUS}\,(I \otimes H) \tag{A.68}$$

follows immediately if we multiply both sides of Eq. (A.67) by $(I \otimes H)$, since $(I \otimes H)^2 = I \otimes I$.

Exercise 3.11 Let us first consider a phase error acting on the target qubit. It transforms the state (3.51) into

$$\left(\alpha\,|0\rangle + \beta\,|1\rangle\right) \otimes \tfrac{1}{\sqrt{2}}\left(|0\rangle - |1\rangle\right) = \tfrac{1}{\sqrt{2}}\left(\alpha\,|00\rangle - \alpha\,|01\rangle + \beta\,|10\rangle - \beta\,|11\rangle\right). \tag{A.69}$$

After application of the CNOT gate, this state becomes

$$\tfrac{1}{\sqrt{2}}\left(\alpha\,|00\rangle - \alpha\,|01\rangle + \beta\,|11\rangle - \beta\,|10\rangle\right) = \tfrac{1}{\sqrt{2}}\left(\alpha\,|0\rangle - \beta|1\rangle\right) \otimes \left(|0\rangle - |1\rangle\right). \tag{A.70}$$

Therefore, this kind of error is particularly dangerous, since, even though initially it only affects the target qubit, after application of the CNOT gate it is also transferred to the control qubit.

Note that this is not the case for the other possible phase or amplitude errors. For instance, a phase error acting on the control qubit transforms the state (3.51) into

$$(\alpha|0\rangle - \beta|1\rangle) \otimes \tfrac{1}{\sqrt{2}} (|0\rangle + |1\rangle),$$

and this state is unchanged by application of the CNOT gate. Therefore, the target qubit is not affected by this phase error. An amplitude error acting on the control qubit transforms (3.51) into

$$(\beta|0\rangle + \alpha|1\rangle) \otimes \tfrac{1}{\sqrt{2}} (|0\rangle + |1\rangle),$$

which is not modified by the CNOT gate. Again, the target qubit is safe. Finally, an amplitude error acting on the target qubit does not affect the computation, that is, the state (3.51) stays the same. This is because the target qubit is symmetric under the amplitude error: $|0\rangle + |1\rangle$ is not modified when $|0\rangle \leftrightarrow |1\rangle$.

Exercise 3.12   By direct computation of the tensor products we obtain:

$$\sigma_1 \otimes \sigma_1 = \begin{bmatrix} 0 & 0 & 0 & 1 \\ 0 & 0 & 1 & 0 \\ 0 & 1 & 0 & 0 \\ 1 & 0 & 0 & 0 \end{bmatrix}, \quad \sigma_1 \otimes \sigma_2 = \begin{bmatrix} 0 & 0 & 0 & -i \\ 0 & 0 & i & 0 \\ 0 & -i & 0 & 0 \\ i & 0 & 0 & 0 \end{bmatrix},$$

$$\sigma_1 \otimes \sigma_3 = \begin{bmatrix} 0 & 0 & 1 & 0 \\ 0 & 0 & 0 & -1 \\ 1 & 0 & 0 & 0 \\ 0 & -1 & 0 & 0 \end{bmatrix}, \quad \sigma_2 \otimes \sigma_1 = \begin{bmatrix} 0 & 0 & 0 & -i \\ 0 & 0 & -i & 0 \\ 0 & i & 0 & 0 \\ i & 0 & 0 & 0 \end{bmatrix},$$

$$\sigma_2 \otimes \sigma_2 = \begin{bmatrix} 0 & 0 & 0 & -1 \\ 0 & 0 & 1 & 0 \\ 0 & 1 & 0 & 0 \\ -1 & 0 & 0 & 0 \end{bmatrix}, \quad \sigma_2 \otimes \sigma_3 = \begin{bmatrix} 0 & 0 & -i & 0 \\ 0 & 0 & 0 & i \\ i & 0 & 0 & 0 \\ 0 & -i & 0 & 0 \end{bmatrix},$$

$$\sigma_3 \otimes \sigma_1 = \begin{bmatrix} 0 & 1 & 0 & 0 \\ 1 & 0 & 0 & 0 \\ 0 & 0 & 0 & -1 \\ 0 & 0 & -1 & 0 \end{bmatrix}, \quad \sigma_3 \otimes \sigma_2 = \begin{bmatrix} 0 & -i & 0 & 0 \\ i & 0 & 0 & 0 \\ 0 & 0 & 0 & i \\ 0 & 0 & -i & 0 \end{bmatrix},$$

$$\sigma_3 \otimes \sigma_3 = \begin{bmatrix} 1 & 0 & 0 & 0 \\ 0 & -1 & 0 & 0 \\ 0 & 0 & -1 & 0 \\ 0 & 0 & 0 & 1 \end{bmatrix}, \quad \sigma_1 \otimes I = \begin{bmatrix} 0 & 0 & 1 & 0 \\ 0 & 0 & 0 & 1 \\ 1 & 0 & 0 & 0 \\ 0 & 1 & 0 & 0 \end{bmatrix},$$

$$\sigma_2 \otimes I = \begin{bmatrix} 0 & 0 & -i & 0 \\ 0 & 0 & 0 & -i \\ i & 0 & 0 & 0 \\ 0 & i & 0 & 0 \end{bmatrix}, \quad \sigma_3 \otimes I = \begin{bmatrix} 1 & 0 & 0 & 0 \\ 0 & 1 & 0 & 0 \\ 0 & 0 & -1 & 0 \\ 0 & 0 & 0 & -1 \end{bmatrix},$$

$$I \otimes \sigma_1 = \begin{bmatrix} 0 & 1 & 0 & 0 \\ 1 & 0 & 0 & 0 \\ 0 & 0 & 0 & 1 \\ 0 & 0 & 1 & 0 \end{bmatrix}, \qquad I \otimes \sigma_2 = \begin{bmatrix} 0 & -i & 0 & 0 \\ i & 0 & 0 & 0 \\ 0 & 0 & 0 & -i \\ 0 & 0 & i & 0 \end{bmatrix},$$

$$I \otimes \sigma_3 = \begin{bmatrix} 1 & 0 & 0 & 0 \\ 0 & -1 & 0 & 0 \\ 0 & 0 & 1 & 0 \\ 0 & 0 & 0 & -1 \end{bmatrix}, \qquad I \otimes I = \begin{bmatrix} 1 & 0 & 0 & 0 \\ 0 & 1 & 0 & 0 \\ 0 & 0 & 1 & 0 \\ 0 & 0 & 0 & 1 \end{bmatrix}. \tag{A.71}$$

Exercise 3.13

$$\langle \psi | \sigma_1 \otimes \sigma_1 | \psi \rangle = \begin{bmatrix} c & \alpha^* & \beta^* & \gamma^* \end{bmatrix} \begin{bmatrix} 0 & 0 & 0 & 1 \\ 0 & 0 & 1 & 0 \\ 0 & 1 & 0 & 0 \\ 1 & 0 & 0 & 0 \end{bmatrix} \begin{bmatrix} c \\ \alpha \\ \beta \\ \gamma \end{bmatrix}$$

$$= c\gamma + c\gamma^* + \alpha\beta^* + \alpha^*\beta,$$

$$\langle \psi | \sigma_1 \otimes \sigma_2 | \psi \rangle = i\left(-c\gamma + c\gamma^* - \alpha\beta^* + \alpha^*\beta\right),$$

$$\langle \psi | \sigma_1 \otimes \sigma_3 | \psi \rangle = c\beta + c\beta^* - \alpha\gamma^* - \alpha^*\gamma,$$

$$\langle \psi | \sigma_2 \otimes \sigma_1 | \psi \rangle = i\left(-c\gamma + c\gamma^* + \alpha\beta^* - \alpha^*\beta\right),$$

$$\langle \psi | \sigma_2 \otimes \sigma_2 | \psi \rangle = -c\gamma - c\gamma^* + \alpha\beta^* + \alpha^*\beta,$$

$$\langle \psi | \sigma_2 \otimes \sigma_3 | \psi \rangle = i\left(-c\beta + c\beta^* - \alpha\gamma^* + \alpha^*\gamma\right),$$

$$\langle \psi | \sigma_3 \otimes \sigma_1 | \psi \rangle = c\alpha + c\alpha^* - \beta\gamma^* - \beta^*\gamma,$$

$$\langle \psi | \sigma_3 \otimes \sigma_2 | \psi \rangle = i\left(-c\alpha + c\alpha^* - \beta\gamma^* + \beta^*\gamma\right),$$

$$\langle \psi | \sigma_3 \otimes \sigma_3 | \psi \rangle = c^2 - |\alpha|^2 - |\beta|^2 + |\gamma|^2,$$

$$\langle \psi | \sigma_1 \otimes I | \psi \rangle = c\beta + \alpha^*\gamma + \beta^*c + \gamma^*\alpha,$$

$$\langle \psi | \sigma_2 \otimes I | \psi \rangle = i\left(-c\beta - \alpha^*\gamma + \beta^*c + \gamma^*\alpha\right),$$

$$\langle \psi | \sigma_3 \otimes I | \psi \rangle = c^2 + |\alpha|^2 - |\beta|^2 - |\gamma|^2,$$

$$\langle \psi | I \otimes \sigma_1 | \psi \rangle = c\alpha + \alpha^*c + \beta^*\gamma + \gamma^*\beta,$$

$$\langle \psi | I \otimes \sigma_2 | \psi \rangle = i\left(-c\alpha + \alpha^*c - \beta^*\gamma + \gamma^*\beta\right),$$

$$\langle \psi | I \otimes \sigma_3 | \psi \rangle = c^2 - |\alpha|^2 + |\beta|^2 - |\gamma|^2,$$

$$\langle \psi | I \otimes I | \psi \rangle = c^2 + |\alpha|^2 + |\beta|^2 + |\gamma|^2 = 1. \tag{A.72}$$

Exercise 3.14 To show that the C^2-U gate can be decomposed as in Fig. 3.10, a tedious but straightforward method is to compute the products of the matrices associated with the gates that build up the decomposition. Since these gates act non-trivially only on two of the three qubits, it is necessary to embed them in the 2^3-dimensional Hilbert space associated with the whole system. The simplest example of embedding is when a 2×2 matrix

$$A = \begin{bmatrix} a & b \\ c & d \end{bmatrix} \tag{A.73}$$

is associated with a linear operator acting on a single qubit and we wish to extend it to the Hilbert space associated with two qubits. We have two possibilities, depending on whether the operator A acts on the less or more significant qubit:

$$I \otimes A = \begin{bmatrix} a & b & 0 & 0 \\ c & d & 0 & 0 \\ 0 & 0 & a & b \\ 0 & 0 & c & d \end{bmatrix}, \qquad A \otimes I = \begin{bmatrix} a & 0 & b & 0 \\ 0 & a & 0 & b \\ c & 0 & d & 0 \\ 0 & c & 0 & d \end{bmatrix}. \tag{A.74}$$

In the present exercise, we must embed 4×4 matrices, associated with operators acting on two qubits, in the Hilbert space for three qubits. Given a generic 4×4 matrix

$$M = \begin{bmatrix} a & b & c & d \\ e & f & g & h \\ i & j & k & l \\ m & n & o & p \end{bmatrix}, \tag{A.75}$$

the embedding gives one of the following three matrices:

$$\begin{bmatrix} a & b & c & d & 0 & 0 & 0 & 0 \\ e & f & g & h & 0 & 0 & 0 & 0 \\ i & j & k & l & 0 & 0 & 0 & 0 \\ m & n & o & p & 0 & 0 & 0 & 0 \\ 0 & 0 & 0 & 0 & a & b & c & d \\ 0 & 0 & 0 & 0 & e & f & g & h \\ 0 & 0 & 0 & 0 & i & j & k & l \\ 0 & 0 & 0 & 0 & m & n & o & p \end{bmatrix}, \begin{bmatrix} a & b & 0 & 0 & c & d & 0 & 0 \\ e & f & 0 & 0 & g & h & 0 & 0 \\ 0 & 0 & a & b & 0 & 0 & c & d \\ 0 & 0 & e & f & 0 & 0 & g & h \\ i & j & 0 & 0 & k & l & 0 & 0 \\ m & n & 0 & 0 & o & p & 0 & 0 \\ 0 & 0 & i & j & 0 & 0 & k & l \\ 0 & 0 & m & n & 0 & 0 & o & p \end{bmatrix}, \begin{bmatrix} a & 0 & b & 0 & c & 0 & d & 0 \\ 0 & a & 0 & b & 0 & c & 0 & d \\ e & 0 & f & 0 & g & 0 & h & 0 \\ 0 & e & 0 & f & 0 & g & 0 & h \\ i & 0 & j & 0 & k & 0 & l & 0 \\ 0 & i & 0 & j & 0 & k & 0 & l \\ m & 0 & n & 0 & o & 0 & p & 0 \\ 0 & m & 0 & n & 0 & o & 0 & p \end{bmatrix},$$

$$\tag{A.76}$$

depending on whether the operator M acts trivially on the first (most significant), second, or third (least significant) qubit.

A much simpler way to solve this exercise is to compute the action of the five gates that build the circuit of Fig. 3.10 on the states $|i_2\, i_1\, i_0\rangle$ of the computational basis and show that their composition is equivalent to the application of the $C^2{-}U$ gate. This way we have

$$|00i_0\rangle \to \quad |00i_0\rangle \quad \to \quad |00i_0\rangle \quad \to \quad |00i_0\rangle \quad \to \quad |00i_0\rangle \quad \to \quad |00i_0\rangle\,,$$

$$|01i_0\rangle \to |01\rangle V|i_0\rangle \to |01\rangle V|i_0\rangle \to \quad |01i_0\rangle \quad \to \quad |01i_0\rangle \quad \to \quad |01i_0\rangle\,,$$

$$|10i_0\rangle \to \quad |10i_0\rangle \quad \to \quad |11i_0\rangle \quad \to |11\rangle V^\dagger|i_0\rangle \to |10\rangle V^\dagger|i_0\rangle \to \quad |10i_0\rangle\,,$$

$$|11i_0\rangle \to |11\rangle V|i_0\rangle \to |10\rangle V|i_0\rangle \to |10\rangle V|i_0\rangle \to |11\rangle V|i_0\rangle \to |11\rangle U|i_0\rangle\,,$$

$$\tag{A.77}$$

where we have used the relations $V^2 = I$ and $V^\dagger V = I = VV^\dagger$. Thus, the circuit of Fig. 3.10 indeed implements the $C^2 - U$ gate, since it acts non-trivially only when the two control qubits are set to 1 and in this case applies a U gate to the third qubit.

Exercise 3.15   We must proceed as in exercise 3.8, but the NOT gates must be applied (before and after a standard $C^{(n-1)}$-NOT gate) to all the qubits that induce non-trivial action of the generalized $C^{(n-1)}$-NOT gate when they are in the state $|0\rangle$.

Exercise 3.16   The matrix D for a 4×4 matrix is given by

$$\begin{bmatrix} \cos \phi_1 & 0 & -\sin \phi_1 & 0 \\ 0 & \cos \phi_2 & 0 & -\sin \phi_2 \\ \sin \phi_1 & 0 & \cos \phi_1 & 0 \\ 0 & \sin \phi_2 & 0 & \cos \phi_2 \end{bmatrix}. \tag{A.78}$$

The circuit in Fig. 3.14 implements the transformation

$$(\overline{\text{CNOT}})\,(R_y(\theta_1) \otimes I)\,(\overline{\text{CNOT}})\,(R_y(\theta_0) \otimes I)\,, \tag{A.79}$$

where

$$R_y(\theta) \otimes I = \begin{bmatrix} \cos \frac{\theta}{2} & 0 & -\sin \frac{\theta}{2} & 0 \\ 0 & \cos \frac{\theta}{2} & 0 & -\sin \frac{\theta}{2} \\ \sin \frac{\theta}{2} & 0 & \cos \frac{\theta}{2} & 0 \\ 0 & \sin \frac{\theta}{2} & 0 & \cos \frac{\theta}{2} \end{bmatrix}, \tag{A.80}$$

and the generalized CNOT gate is given by

$$\overline{\text{CNOT}} = \begin{bmatrix} 1 & 0 & 0 & 0 \\ 0 & 0 & 0 & 1 \\ 0 & 0 & 1 & 0 \\ 0 & 1 & 0 & 0 \end{bmatrix}. \tag{A.81}$$

It is easy to check the equality between (A.78) and (A.79) by direct matrix multiplication, provided that $\theta_0 = \phi_1 + \phi_2$ and $\theta_1 = \phi_1 - \phi_2$.

Exercise 3.17   We have

$$U_f^2 |x\rangle|y\rangle \;=\; U_f |x\rangle|y \oplus f(x)\rangle \;=\; |x\rangle|y \oplus 2f(x)\rangle \;=\; |x\rangle|y\rangle\,, \tag{A.82}$$

where $|x\rangle \equiv |x_{n-1}, x_{n-2}, \ldots, x_0\rangle$. Hence, $U_f^2 = I$, that is,

$$U_f^{-1} \;=\; U_f\,. \tag{A.83}$$

Let us show that U_f is Hermitian. The matrix elements of U_f in the computational basis are given by

$$\begin{aligned} U_f(x, y; x', y') &= \langle x|\langle y|U_f|x'\rangle|y'\rangle \\ &= \langle x|x'\rangle\langle y|y' \oplus f(x')\rangle = \delta_{x,x'}\delta_{y,y' \oplus f(x)}\,. \end{aligned} \tag{A.84}$$

We now compute the matrix elements of the adjoint operator $U_f^\dagger$:

$$U_f^\dagger(x, y; x', y') = U_f^\star(x', y'; x, y) = U_f(x', y'; x, y) = \delta_{x,x'}\delta_{y',y\oplus f(x)}. \quad \text{(A.85)}$$

Since $y = y' \oplus f(x)$ can be written equivalently as $y \oplus f(x) = y' \oplus 2f(x) = y'$, it follows that

$$U_f^\dagger(x, y; x', y') = U_f(x, y; x', y'). \quad \text{(A.86)}$$

Eqs. (A.83) and (A.86) imply that

$$U_f^\dagger = U_f^{-1}, \quad \text{(A.87)}$$

that is, U_f is unitary.

Exercise 3.18 It is clear from Eq. (3.132) that the probability that Grover's algorithm fails is given by

$$p(x \neq x_0) = \cos^2[(2k+1)\theta], \quad \text{(A.88)}$$

where $\theta \approx 1/\sqrt{N}$ and

$$(2k+1)\theta = \frac{\pi}{2} + O(\theta) = \frac{\pi}{2} + O(\sqrt{N}). \quad \text{(A.89)}$$

It follows that

$$p(x \neq x_0) = \cos^2\left[\frac{\pi}{2} + O\left(\frac{1}{\sqrt{N}}\right)\right] = O\left(\frac{1}{N}\right). \quad \text{(A.90)}$$

Exercise 3.19 One step of Grover's algorithm requires an oracle query, $2n$ Hadamard gates and a reflection about the hyperplane orthogonal to $|0\rangle$. To operate this reflection, we need to implement a generalized $C^{(n-1)}$–MINUS gate, which puts a minus sign in front of the state vector $|00 \cdots 0\rangle$. As we saw in Sec. 3.5 (see Fig. 3.11), this transformation can be decomposed into $2(n-2)$ Toffoli gates plus a single CMINUS gate. The price to pay is that $n-2$ ancillary qubits are required. Alternatively, it is possible to compute the $C^{(n-1)}$–MINUS gate without ancillary qubits in $O(n^2)$ elementary gates (see Barenco *et al.*, 1995). Finally, we assume that the oracle answers the query instantaneously, that is, the time it takes to operate is not included in the complexity analysis. This is because the cost of the oracle call depends upon the specific application. In conclusion, one step of Grover's algorithm takes a time (measured in the number of elementary gates) of the order of n (with ancillary qubits) or of the order of n^2 (without ancillaries). Since Grover's algorithm requires $O(\sqrt{N})$ steps, in both cases we need $O(\sqrt{N}\log N)$ elementary gates.

Exercise 3.20 Let us call $|\psi\rangle$ the exact wave function at the end of the quantum Fourier transform and $|\tilde{\psi}\rangle$ the actual wave function when unitary errors take place. We know from Sec. 3.6 that, if errors are uniformly bound at each step by some constant δ, then

$$\left\| |\tilde{\psi}\rangle - |\psi\rangle \right\| < n_g \, \delta \,, \tag{A.91}$$

where $n_g = O(n^2)$ is the number of elementary quantum gates required to implement the Fourier transform. If the desired accuracy in the output state is ϵ, it is sufficient to implement the single quantum gates with precision δ such that $n_g \delta < \epsilon$. Therefore, $\delta = O(1/n^2)$, namely, it drops only polynomially with the number of qubits. An interesting consequence is the following (Coppersmith, 1994). If a final accuracy ϵ is required, we can simply skip from the quantum Fourier transform algorithm the controlled-phase shift gates R_k of angles $2\pi i/2^k < \epsilon/n_g$. This is just because these R_k gates differ from the identity by less than ϵ/n_g. This observation ensures that it is not necessary to perform controlled-phase shift gates with exponentially small phases; a polynomial control of the phases is sufficient.

Exercise 3.21 Since $F^{-1}F = I$, it is sufficient to run the circuit in Fig. 3.25 from right to left.

Exercise 3.22 If we write the state of the qubit at time t as

$$|\psi(t)\rangle = \alpha(t)|0\rangle + \beta(t)|1\rangle \,, \tag{A.92}$$

then equation (3.22) can be expressed as

$$i\hbar \frac{d}{dt} \begin{bmatrix} \alpha(t) \\ \beta(t) \end{bmatrix} = -\mu \begin{bmatrix} H_0 & H_1 \, e^{-i\omega t} \\ H_1 \, e^{i\omega t} & -H_0 \end{bmatrix} \begin{bmatrix} \alpha(t) \\ \beta(t) \end{bmatrix} . \tag{A.93}$$

If we define $\omega_0 \equiv -2\mu H_0/\hbar$ and $\omega_1 \equiv -2\mu H_1/\hbar$, the we can write the Schrödinger equation (A.93) in the form

$$\begin{cases} i\dfrac{d}{dt}\,\alpha(t) = \dfrac{\omega_0}{2}\,\alpha(t) + \dfrac{\omega_1}{2}\,e^{-i\omega t}\beta(t)\,, \\[2mm] i\dfrac{d}{dt}\,\beta(t) = \dfrac{\omega_1}{2}\,e^{i\omega t}\alpha(t) - \dfrac{\omega_0}{2}\,\beta(t)\,. \end{cases} \tag{A.94}$$

Equations (A.94) constitute a linear homogeneous system with time-dependent coefficients. To solve this system, it is convenient to define new functions

$$a(t) \equiv \alpha(t)\,\exp(i\omega t/2) \qquad \text{and} \qquad b(t) \equiv \beta(t)\,\exp(-i\omega t/2)\,. \tag{A.95}$$

If we introduce the vector

$$|\tilde{\psi}(t)\rangle = a(t)\,|0\rangle + b(t)\,|1\rangle \,, \tag{A.96}$$

we can easily see that

$$|\tilde{\psi}(t)\rangle = R_z(-\omega t)\,|\psi(t)\rangle \,, \tag{A.97}$$

where the rotation matrix $R_z(-\omega t)$ was defined in Eq. (3.29) and represents a rotation of the Bloch sphere through an angle $-\omega t$ about the axis z. Therefore, the transformation (A.95) corresponds to the change from a fixed reference frame to a frame rotating with the frequency ω of the oscillating magnetic field. Substituting (A.95) into (A.94), we obtain

$$\begin{cases} i\dfrac{d}{dt}\,a(t) = \left(\dfrac{\omega_0 - \omega}{2}\right) a(t) + \dfrac{\omega_1}{2}\,b(t)\,, \\[2mm] i\dfrac{d}{dt}\,b(t) = \dfrac{\omega_1}{2}\,a(t) - \left(\dfrac{\omega_0 - \omega}{2}\right) b(t)\,. \end{cases} \tag{A.98}$$

Note that this new system of equations has constant coefficients. It corresponds to the Schrödinger equation in the rotating frame and can also be written as

$$i\hbar \frac{d}{dt}\,|\tilde{\psi}(t)\rangle = \tilde{H}\,|\tilde{\psi}(t)\rangle\,, \tag{A.99}$$

where the Hamiltonian

$$\tilde{H} = \frac{\hbar}{2} \begin{bmatrix} \omega_0 - \omega & \omega_1 \\ \omega_1 & -(\omega_0 - \omega) \end{bmatrix} \tag{A.100}$$

is time-independent. Thus, we obtain

$$|\tilde{\psi}(t)\rangle = \tilde{U}\,|\tilde{\psi}(0)\rangle = \tilde{U}\,|\psi(0)\rangle\,, \tag{A.101}$$

where the unitary time-evolution operator $\tilde{U}$ is given by

$$\tilde{U} = e^{-i\tilde{H}t/\hbar} = e^{-i[(\omega_0 - \omega)\sigma_z + \omega_1 \sigma_x]t/2}\,. \tag{A.102}$$

Finally, we obtain the formal solution to the Schrödinger equation (3.22):

$$|\psi(t)\rangle = R_z(\omega t)\,|\tilde{\psi}(t)\rangle = e^{-i\omega\sigma_z t/2} e^{-i[(\omega_0 - \omega)\sigma_z + \omega_1 \sigma_x]t/2}\,|\psi(0)\rangle\,. \tag{A.103}$$

Since the Hamiltonian $\tilde{H}$ does not explicitly depend on time, we can write the solution of (3.22) in the form (2.140):

$$|\psi(t)\rangle = \sum_{n=1}^{2} c_n(0) \exp\left(-\frac{i}{\hbar}E_n t\right) |n\rangle\,. \tag{A.104}$$

Here E_1 and E_2 denote the eigenvalues of the Hamiltonian $\tilde{H}$ and the coefficients $c_n(0)$ $(n = 1, 2)$ are given by

$$c_n(0) = \langle\varphi_n|\tilde{\psi}(0)\rangle = \langle\varphi_n|\psi(0)\rangle\,, \tag{A.105}$$

$|\varphi_1\rangle$ and $|\varphi_2\rangle$ being the eigenvectors of $\tilde{H}$. It is easy to find that

$$E_1 = \frac{\hbar}{2}\sqrt{(\omega_0 - \omega)^2 + \omega_1^2}, \qquad E_2 = -\frac{\hbar}{2}\sqrt{(\omega_0 - \omega)^2 + \omega_1^2}, \qquad (A.106)$$

$$|\varphi_1\rangle = \begin{bmatrix} \cos\frac{\theta}{2} \\ \sin\frac{\theta}{2} \end{bmatrix}, \qquad |\varphi_2\rangle = \begin{bmatrix} -\sin\frac{\theta}{2} \\ \cos\frac{\theta}{2} \end{bmatrix}, \qquad (A.107)$$

where the angle θ is defined by

$$\tan\theta = \frac{\omega_1}{\omega_0 - \omega_1}. \qquad (A.108)$$

Let us assume that at time $t = 0$ the system is in the state

$$|\psi(0)\rangle = |\tilde{\psi}(0)\rangle = |0\rangle. \qquad (A.109)$$

This corresponds to

$$c_1(0) = \langle\varphi_1|0\rangle = \cos\frac{\theta}{2}, \qquad c_2(0) = \langle\varphi_2|0\rangle = -\sin\frac{\theta}{2}. \qquad (A.110)$$

After substitution of (A.110) into the general solution (A.104), we obtain

$$|\tilde{\psi}(t)\rangle = \cos\frac{\theta}{2}\, e^{-iE_1 t/\hbar}\, |\varphi_1\rangle - \sin\frac{\theta}{2}\, e^{-iE_2 t/\hbar}\, |\varphi_2\rangle. \qquad (A.111)$$

We can now compute the probability $p_1(t)$ of finding the spin-$\frac{1}{2}$ particle in the state $|1\rangle$ at time t. We obtain

$$p_1(t) = \left|\langle 1|\psi(t)\rangle\right|^2 = \left|\beta(t)\right|^2 = \left|b(t)\right|^2 = \left|\langle 1|\tilde{\psi}(t)\rangle\right|^2$$

$$= \frac{\omega_1^2}{(\omega_0 - \omega)^2 + \omega_1^2} \sin^2\left(\sqrt{(\omega_0 - \omega)^2 + \omega_1^2}\,\frac{t}{2}\right). \qquad (A.112)$$

This probability is equal to 0 when $t = 0$ and varies sinusoidally between 0 and $\omega_1^2/[(\omega_0 - \omega)^2 + \omega_1^2]$. These oscillations take place with frequency $\Omega = \sqrt{(\omega_0 - \omega)^2 + \omega_1^2}$. Formula (A.112) is known as *Rabi's formula* and Ω is the Rabi frequency. The resonant case $\omega = \omega_0$ is particularly important. In this case, the state of the particle oscillates with frequency $\Omega = \omega_1$ between the states $|0\rangle$ and $|1\rangle$. We have $p_1(t) = 1$ at times $t = (2n + 1)\pi/\omega_1$. Note that far from resonance the transition probability between the states $|0\rangle$ and $|1\rangle$ remains small, that is, the probability to measure the z-component of the spin and obtain $\sigma_z = -1$ is small at all times.

Chapter 4

Exercise 4.1   Let f_A and f_B denote the public encryption keys whose corresponding secret decryption keys f_A^{-1} and f_B^{-1} are possessed by Alice alone and by Bob alone, respectively. Alice encrypts her signature S_A by means of her secret decryption key f_A^{-1} into $f_A^{-1}(S_A)$. She then encrypts the plain text P plus her signature using Bob's public encryption key. This way she produces the cypher text $C = f_B(P + f_A^{-1}(S_A))$, which is sent over a public channel to Bob. Bob then decrypts the cypher text C by means of his secret key f_B^{-1} as follows: $f_B^{-1}(C) = P + f_A^{-1}(S_A)$. After this he uses the public key f_A to verify Alice's signature $S_A = f_A(f_A^{-1}(S_A))$. We underline that the above authentication procedure is valid because only Alice knows her secret key f_A^{-1}: nobody else could have produced $f_A^{-1}(S_A)$.

Exercise 4.2   The average fidelity is given by

$$\overline{F} = \frac{1}{4\pi} \int_0^{2\pi} d\phi \int_0^{\pi} d\theta \, \sin\theta \, |\langle \psi_1 | \psi_2 \rangle|^2 \,, \qquad (A.113)$$

where the spherical coordinates θ and ϕ single out the state $|\psi_1\rangle$ on the Bloch sphere (see Sec. 2.1). To compute the integral (A.113) it is convenient to choose the z-axis along the polarization direction of $|\psi_2\rangle$. This choice corresponds to $|\psi_2\rangle = |0\rangle$ and therefore we have

$$\overline{F} = \frac{1}{4\pi} \int_0^{2\pi} d\phi \int_0^{\pi} d\theta \, \sin\theta \, \cos^2 \tfrac{\theta}{2} = \tfrac{1}{2} \,. \qquad (A.114)$$

Exercise 4.3   Alice and Bob share the EPR state $|\phi^+\rangle = \frac{1}{\sqrt{2}}(|00\rangle + |11\rangle)$. If Alice measures σ_x, she obtains outcomes ± 1 with equal probabilities $p_+^{(A)} = p_-^{(A)} = \frac{1}{2}$. After Alice's measurement the state of Bob's qubit is

$$|\psi_\pm\rangle_B = \frac{1}{\sqrt{2}}(|0\rangle \pm |1\rangle)\,, \qquad (A.115)$$

where the $\pm$ sign corresponds to the ± 1 result of Alice's measurement. If Bob performs a spin measurement along the u axis singled out by the spherical coordinates θ and ϕ, he obtains $\sigma_u = +1$ with probability

$$\left| {}_u\langle + | \psi_\pm \rangle_B \right|^2 = \tfrac{1}{2}\left[1 \pm \cos\phi \, \sin\theta \right]\,, \qquad (A.116)$$

where $|+\rangle_u$ is the eigenvector of the operator σ_u corresponding to the eigenvalue $+1$ (the explicit expressions for σ_u and $|+\rangle_u$ are given by Eq. (2.159) and Eq. (2.160), respectively). The eigenvector $|-\rangle_u$ of σ_u corresponding to the eigenvalue -1 is obtained from $|+\rangle_u$ via the transformation $\phi \to \phi + \pi$ and $\theta \to \pi - \theta$. Hence, Bob obtains $\sigma_u = -1$ with probability

$$\left| {}_u\langle - | \psi_\pm \rangle_B \right|^2 = \tfrac{1}{2}\left[1 \pm \cos(\phi + \pi) \, \sin(\pi - \theta) \right] = \tfrac{1}{2}\left[1 \mp \cos\phi \, \sin\theta \right]\,. \quad (A.117)$$

Since Bob receives the states $|\psi_\pm\rangle_B$ with probabilities $\frac{1}{2}$, then he obtains $\sigma_u = +1$ with probability

$$p_+^{(B)} = \frac{1}{2}\left|_u\langle+|\psi_+\rangle_B\right|^2 + \frac{1}{2}\left|_u\langle+|\psi_-\rangle_B\right|^2 = \frac{1}{2}, \qquad (A.118a)$$

and $\sigma_u = -1$ with probability

$$p_-^{(B)} = \frac{1}{2}\left|_u\langle-|\psi_+\rangle_B\right|^2 + \frac{1}{2}\left|_u\langle-|\psi_-\rangle_B\right|^2 = \frac{1}{2}. \qquad (A.118b)$$

If instead Alice measures σ_z, then the state of Bob's qubit collapses onto $|\phi_+\rangle_B = |0\rangle$ or $|\phi_-\rangle_B = |1\rangle$ with probabilities $\frac{1}{2}$. Since

$$\left|_u\langle+|\phi_+\rangle_B\right|^2 = \cos^2\frac{\theta}{2}, \qquad \left|_u\langle+|\phi_-\rangle_B\right|^2 = \sin^2\frac{\theta}{2},$$
$$\left|_u\langle-|\phi_+\rangle_B\right|^2 = \sin^2\frac{\theta}{2}, \qquad \left|_u\langle-|\phi_-\rangle_B\right|^2 = \cos^2\frac{\theta}{2}, \qquad (A.119)$$

then Bob's spin measurement along the u axis gives outcomes ±1 with probabilities

$$p_+^{(B)} = \frac{1}{2}, \qquad p_-^{(B)} = \frac{1}{2}. \qquad (A.120)$$

This result is trivially independent of the chosen axis u. It is important to note that the outcomes $p_+^{(B)}$ and $p_-^{(B)}$ always have the same probability. This implies that the EPR phenomenon cannot be used for faster than light communication. Whatever axis Alice and Bob choose for their measurements, Bob always obtains randomly $+1$ or -1. Therefore, no information has been transmitted from Alice to Bob.

Exercise 4.4 We must consider only the cases in which Alice and Bob used the same alphabet, since the raw key shared by Alice and Bob comes from here. We have the following four cases:

Case	1	2	3	4
Alice's data bits	0	0	1	1
Alphabet	x	z	x	z
Transmitted qubits	$|+\rangle$	$|0\rangle$	$|-\rangle$	$|1\rangle$

Eve measures the spin polarization σ_u along an arbitrary direction singled out by the spherical coordinates θ and ϕ (see Sec. 1.5). She obtains one bit of information as follows: if the outcome of her measurement is $\sigma_u = +1$, she decides that the bit value is 0; if the outcome is $\sigma_u = -1$, the bit is 1. The probabilities of these two outcomes are given by

$$p_0^{(i)} = \left|_u\langle+|\psi^{(i)}\rangle\right|^2, \qquad p_1^{(i)} = \left|_u\langle-|\psi^{(i)}\rangle\right|^2, \qquad (A.121)$$

where $|+\rangle_u$ and $|-\rangle_u$ are the eigenstates of σ_u corresponding to the eigenvalues $+1$ and -1 (their explicit expressions are given by Eq. (2.160)), and the index (i) denotes one of the four possible transmitted qubits, that is, $|\psi^{(1)}\rangle = |+\rangle$,

$|\psi^{(2)}\rangle = |0\rangle$, $|\psi^{(3)}\rangle = |-\rangle$ and $|\psi^{(4)}\rangle = |1\rangle$. We have eight possibilities ($p_0^{(i)}$ and $p_1^{(i)}$, with $i = 1, \ldots, 4$), which take place with the following probabilities:

$$
\begin{aligned}
p_0^{(1)} &= \tfrac{1}{2}\left(1 + \sin\theta\right) = p_1^{(3)}, & p_1^{(1)} &= \tfrac{1}{2}\left(1 - \sin\theta\right) = p_0^{(3)}, \\
p_0^{(2)} &= \cos^2\tfrac{\theta}{2} = p_1^{(4)}, & p_1^{(2)} &= \sin^2\tfrac{\theta}{2} = p_0^{(4)}.
\end{aligned} \tag{A.122}
$$

After her measurement, Eve resends the state $|+\rangle_u$ or the state $|-\rangle_u$ with the probabilities $p_0^{(i)}$ and $p_1^{(i)}$, respectively. Bob measures this state in the same basis as Alice's original basis (remember that we are interested in the bits that constitute the raw key). There are sixteen possible cases, corresponding to the state obtained by Bob, provided that Alice sent a given state and Eve resent another state. The error rate is obtained by adding all the cases in which the bit obtained by Bob differs from the original Alice's bit. Using this procedure, it is easy to check that the error rate is $\tfrac{1}{4}$.

Exercise 4.5   In order to measure the state $|\psi_1\rangle$ without disturbing it, we must measure an observable such that $|\psi_1\rangle$ is an eigenstate of the Hermitian operator associated with the observable. As we wish to gain information about which state was sent by Alice without disturbing the state, we require that $|\psi_1\rangle$ and $|\psi_2\rangle$ both be eigenstates of the same Hermitian operator, corresponding to two different eigenvalues. However, these requirements cannot be fulfilled, because we have assumed that the states $|\psi_1\rangle$ and $|\psi_2\rangle$ are not orthogonal. Therefore, $|\psi_1\rangle$ and $|\psi_2\rangle$ cannot be eigenstates of the same operator and any measurement necessarily disturbs at least one of the two states.

Exercise 4.6   We start from the initial state $|\psi\rangle|0\rangle|0\rangle$, where $|\psi\rangle = \alpha|0\rangle + \beta|1\rangle$, and compute the quantum gates represented in Fig. 4.8. Let us write down explicitly the first few steps and the final result:

$$
\begin{aligned}
|\psi\rangle|0\rangle|0\rangle &\rightarrow |\psi\rangle\tfrac{1}{\sqrt{2}}\left(|0\rangle + |1\rangle\right)|0\rangle \rightarrow \tfrac{1}{\sqrt{2}}\left(\alpha|0\rangle + \beta|1\rangle\right)\left(|00\rangle + |11\rangle\right) \\
&\rightarrow \tfrac{1}{\sqrt{2}}\left(\alpha|000\rangle + \alpha|011\rangle + \beta|110\rangle + \beta|101\rangle\right) \\
&\rightarrow \ldots \\
&\rightarrow \tfrac{1}{2}\left(|00\rangle + |01\rangle + |10\rangle + |11\rangle\right)\left(\alpha|0\rangle + \beta|1\rangle\right). \tag{A.123}
\end{aligned}
$$

Given the final state of (A.123), it is also possible to apply two Hadamard gates to the first two qubits and end up with the state $|0\rangle|0\rangle|\psi\rangle$, which coincides with the initial state, except for a permutation of the qubit states.

Exercise 4.7   Similarly, to the previous exercise, we write down the action of the first few quantum gates of the circuit drawn in Fig. 4.9 and the final result:

$$\frac{1}{\sqrt{2}} \left(\alpha \left| 01 \right\rangle + \beta \left| 10 \right\rangle \right) \left(\left| 000 \right\rangle + \left| 111 \right\rangle \right)$$

$$\rightarrow \frac{1}{2} \left(\alpha \left| 00 \right\rangle - \alpha \left| 01 \right\rangle + \beta \left| 10 \right\rangle + \beta \left| 11 \right\rangle \right) \left(\left| 000 \right\rangle + \left| 111 \right\rangle \right)$$

$$\rightarrow \frac{1}{2} \left(\alpha \left| 00000 \right\rangle + \alpha \left| 00111 \right\rangle - \alpha \left| 01100 \right\rangle - \alpha \left| 01011 \right\rangle \right.$$
$$\left. + \beta \left| 10000 \right\rangle + \beta \left| 10111 \right\rangle + \beta \left| 11100 \right\rangle + \beta \left| 11011 \right\rangle \right)$$

$$\rightarrow \ldots$$

$$\rightarrow \frac{1}{2} \left| 1 \right\rangle \left(\left| 00 \right\rangle + \left| 01 \right\rangle + \left| 10 \right\rangle + \left| 11 \right\rangle \right) \left(\alpha \left| 01 \right\rangle + \beta \left| 10 \right\rangle \right) . \tag{A.124}$$

Bibliography

Abrams, D. S. and Lloyd, S. (1997), Simulation of many-body Fermi systems on a universal quantum computer, *Phys. Rev. Lett.* **79**, 2586.

Abrams, D. S. and Lloyd, S. (1999), Quantum algorithm providing exponential speed increase for finding eigenvalues and eigenvectors, *Phys. Rev. Lett.* **83**, 5162.

Aharonov, D. (2001), Lecture notes for quantum computing, available at http://www.cs.huji.ac.il/~doria/.

Agaian, S. S. and Klappenecker, A. (2002), Quantum computing and a unified approach to fast unitary transforms, quant-ph/0201120.

Alber, G., Beth, T., Horodecki, M., Horodecki, P., Horodecki, R., Rötteler, M., Weinfurter, H., Werner, R., and Zeilinger, A. (2001), Quantum information – An introduction to theoretical concepts and experiments, Springer–Verlag.

Alekseev, V. M. and Jacobson, M. V. (1981), Symbolic dynamics and hyperbolic dynamic systems, *Phys. Rep.* **75**, 287.

Aspect, A., Grangier, P., and Roger, G. (1981), Experimental tests of realistic local theories via Bell's theorem, *Phys. Rev. Lett.* **47**, 460.

Asplemeyer, M., Böhm, H. R., Gyatso, T., Jennewein, T., Kaltenbaek, R., Lindenthal, M., Molina-Terriza, G., Poppe, A., Resch, K., Taraba, M., Ursin, R., Walther, P., and Zeilinger, A. (2003), Long-distance free-space distribution of quantum entanglement, *Science* **301**, 621.

Balazs, N. L. and Voros, A. (1989), The quantized baker's transformation, *Ann. Phys.* (N.Y.) **190**, 1.

Barenco, A. (1995), A universal two-bit gate for quantum computation, *Proc. R. Soc. Lond.* A **449**, 679.

Barenco, A., Bennett, C. H., Cleve, R., DiVincenzo, D. P., Margolus, N., Shor, P., Sleator, T., Smolin, J. A., and Weinfurter, H. (1995), Elementary gates for quantum computation, *Phys. Rev. A* **52**, 3457.

Beauregard, S. (2003), Circuit for Shor's algorithm using 2n+3 qubits, *Quantum Computation and Information* **3**, 175.

Beckman, B., Chari A. N., Devabhaktuni, S., and Preskill, J. (1996), Efficient networks for quantum factoring, *Phys. Rev. A* **54**, 1034.

Bell, J. S. (1964), On the Einstein–Podolsky–Rosen paradox, *Physics* **1**, 195.

Benenti, G., Casati, G., Montangero, S., and Shepelyansky, D. L. (2001), Efficient quantum computing of complex dynamics, *Phys. Rev. Lett.* **87**, 227901.

Benenti, G., Casati, G., Montangero, S., and Shepelyansky, D. L. (2003), Dynamical localization simulated on a few-qubit quantum computer, *Phys. Rev. A* **67**, 052312.

Bennett, C. H. (1973), Logical reversibility of computation, *IBM J. Res. Dev.* **17**, 525.

Bennett, C. H. (1982), The thermodynamics of computation – a review, *Int. J. Theor. Phys.* **21**, 905.

Bennett, C. H. and Brassard, G. (1984), Quantum cryptography: Public key distribution and coin tossing, in Proc. of IEEE Int. Conf. on Computers, Systems and Signal Processing, p. 175, IEEE, New York, 1984.

Bennett, C. H. (1987), Demons, engines and the second law, *Sci. Am.* **257:5**, 108, November 1987.

Bennett, C. H., Brassard, G., and Ekert, A. K. (1991), Quantum cryptography, *Sci. Am.*, **267:4**, 50, October 1992.

Bennett, C. H., Brassard, G., and Mermin, N. D. (1992), Quantum cryptography without Bell's theorem, *Phys. Rev. Lett.* **68**, 557.

Bennett, C. H. (1992), Quantum cryptography using any two nonorthogonal states, *Phys. Rev. Lett.* **68**, 3121.

Bennett, C. H. and Wiesner, S. J. (1992), Communication via one- and two-particle operators on Einstein–Podolsky–Rosen states, *Phys. Rev. Lett.* **69**, 2881.

Bennett, C. H., Brassard, G., Crépau, C., Jozsa, R., Peres, A., and Wootters, W. K. (1993), Teleporting an unknown quantum state via dual classical and Einstein–Podolsky–Rosen channels, *Phys. Rev. Lett.* **70**, 1895.

Bennett, C. H. and DiVincenzo, D. P. (2000), Quantum information and computation, *Nature* **404**, 247.

Bernstein, E. and Vazirani, U. (1997), Quantum complexity theory, *SIAM J. Comput.* **26**, 1411.

Biham, E., Brassard, G., Kenigsberg, D., and Mor, T. (2003), Quantum computing without entanglement, quant-ph/0306182.

Bohr, N. (1935), Can quantum-mechanical description of physical reality be considered complete?, *Phys. Rev.* **48**, 696.

Boschi, D., Branca, S., De Martini, F., Hardy, L., and Popescu, S. (1998), Experimental realization of teleporting an unknown pure quantum state via dual classical and Einstein–Podolsky–Rosen channels, *Phys. Rev. Lett.* **80**, 1121.

Bouwmeester, D., Pan, J.-W., Mattle, K., Eibl, M., Weinfurter, H., and Zeilinger, A. (1997), Experimental quantum teleportation, *Nature* **390**, 575.

Bouwmeester, D., Ekert. A., and Zeilinger, A. (Eds.) (2000), The Physics of Quantum Information, Springer–Verlag.

Bowden, C. M., Chen, G., Diao, Z., and Klappenecker, A. (2000), The universality of the quantum Fourier transform in forming the basis of quantum computing algorithms, quant-ph/0007122.

Boyer, B., Brassard, G., Høyer, P., and Tapp, A. (1998), Tight bounds on quantum searching, *Fortschr. Phys.* **46**, 493.

Brassard, G., Braunstein, S. L., and Cleve, R. (1998), Teleportation as a quantum computation, *Physica* **D120**, 43.

Brassard, G., Høyer, P., Mosca, M., and Tapp, A. (2002), Quantum amplitude amplification and estimation, quant-ph/0005055, in *Contemporary Mathematics* **305**, AMS Special Session: Quantum Computation and Information, Lomonaco, S. J. and Brandt, H. E. (Eds.), American Mathematical Society, Providence, RI.

Briegel, H.-J., Dür, W., Cirac, J. I., and Zoller, P. (1998), Quantum repeaters: The role of imperfect local operations in quantum communication, *Phys. Rev. Lett.* **81**, 5932.

Brylinski, R. K. and Chen, G. (Eds.) (2002), Mathematics of quantum computation, Chapman & Hall/CRC.

Bruß, D., DiVincenzo, D. P., Ekert, A., Fuchs, C. A., Macchiavello, C., and Smolin, J. A. (1998), Optimal universal and state-dependent quantum cloning, *Phys. Rev. A* **57**, 2368.

Bruß, D. and Lütkenhaus, N. (2000), Quantum key distribution: from principles to practicalities, quant-ph/9901061, in *Appl. Algebra Eng. Commun. Comput. (AAECC)* **10**, 383.

Buttler, W. T., Hughes, R. J., Lamoreaux, S. K., Morgan, G. L., Nordholt, J. E., and Peterson, C. G. (2000), Daylight quantum key distribution over 1.6 km, *Phys. Rev. Lett.* **84**, 5652.

Bužek, V. and Hillery, M. (1996), Quantum copying: Beyond the no-cloning theorem, *Phys. Rev.* **A54**, 1844; Universal optimal cloning of qubits and quantum registers, quant-ph/9801009.

Cabello, A. (2000–2003), Bibliographic guide to the foundations of quantum mechanics and quantum information, quant-ph/0012089.

Casati, G. and Chirikov, B. V. (Eds.) (1995), Quantum chaos: between order and disorder, Cambridge University Press, Cambridge.

Chiorescu, I., Nakamura, Y., Harmans, C. J. P. M., and Mooij, J. E. (2003), Coherent quantum dynamics of a superconducting flux qubit, *Science* **299**, 1869.

Cirac, J. I. and Zoller, P. (1995), Quantum computations with cold trapped ions, *Phys. Rev. Lett.* **74**, 4091.

Cirac, J. I. and Zoller, P. (2001), A scalable quantum computer with ions in an array of microtraps, *Nature* **404**, 579.

Clauser, J. F., Horne, M. A., Shimony, A., and Holt, R. A., (1969), Proposed experiment to test local hidden-variable theories, *Phys. Rev. Lett.* **23**, 880.

Cleve, R., Ekert, A., Macchiavello, C., and Mosca, M. (1998), Quantum algorithms revisited, *Proc. R. Soc. Lond. A* **454**, 339.

Church, A. (1936), An unsolvable problem of elementary number theory, *Am. J. Math.* **58**, 345.

Cohen-Tannoudji, C., Diu, B., and Laloë, F. (1977), Quantum mechanics, vols. I and II, Hermann, Paris.

Coppersmith, D. (1994), An approximate Fourier transform useful in quantum factoring, IBM Research Report No. RC 19642, quant-ph/0201067.

Cormen, T. H., Leiserson, C. E., Rivest, R. L., and Stein, C. (2001), Introduction to algorithms, MIT Press, Cambridge, Massachusetts.

Dana, I., Murray, N. W., and Percival, I. C. (1989), Resonances and diffusion in periodic Hamiltonian maps, *Phys. Rev. Lett.* **62**, 233.

Deutsch, D. (1985), Quantum theory, the Church–Turing principle and the universal quantum computer, *Proc. R. Soc. Lond. A* **400**, 97.

Deutsch, D. (1989), Quantum computational networks, *Proc. R. Soc. Lond. A* **425**, 73.

Deutsch, D. and Jozsa, R. (1992), Rapid solution of problems by quantum computation, *Proc. R. Soc. Lond. A* **439**, 553.

Deutsch, D., Barenco, A., and Ekert, A. (1995), Universality in quantum computation, *Proc. R. Soc. Lond. A* **449**, 669.

Dieks, D. (1982), Communication by EPR devices, *Phys. Lett. A* **92**, 271.

DiVincenzo, D. P. (1995), Two-bit gates are universal for quantum computation, *Phys. Rev. A* **51**, 1015.

DiVincenzo, D. P. (2000), The physical implementation of quantum computation, *Fortschr. Phys.* **48**, 771.

Draper, T. G. (2000), Addition on a quantum computer, quant-ph/0008033.

Einstein, A., Podolsky, B., and Rosen, N. (1935), Can quantum-mechanical description of physical reality be considered complete?, *Phys. Rev.* **47**, 777.

Ekert, A. K. (1991), Quantum cryptography based on Bell's theorem, *Phys. Rev. Lett.* **67**, 661.

Ekert, A. and Jozsa, R. (1996), Quantum computation and Shor's factoring algorithm, *Rev. Mod. Phys.* **68**, 733.

Ekert, A. and Jozsa, R. (1998), Quantum algorithms: Entanglement enhanced information processing, *Phil. Trans. R. Soc. Lond. A* **356**, 1769.

Ekert, A., Hayden, P.M., and Inamori, H. (2001), Basic concepts in quantum computation, quant-ph/0011013, in "Coherent atomic matter waves", Les Houches Summer Schools, Session LXXII, Kaiser, R., Westbrook, C., and David, F. (Eds.), Springer–Verlag.

Emerson, J., Weinstein, Y. S., Lloyd, S., and Cory, D. G. (2002), Fidelity decay as an efficient indicator of quantum chaos, *Phys. Rev. Lett.* **89**, 284102.

Emerson, J., Lloyd, S., Poulin, D., and Cory, D. G. (2003), Estimation of the local density of states on a quantum computer, quant-ph/0308164.

Feynman, R. P. (1982), Simulating Physics with Computers, *Int. J. Theor. Phys.* **21**, 467.

Feynman, R. P. (1996), Feynman lectures on computation, Hey, T. and Allen, R. W. (Eds.), Perseus Publishing.

Fijany, A. and Williams, C. P. (1998), Quantum wavelet transforms: fast algorithms and complete circuits, quant-ph/9809004, in Lecture Notes in Computer Science, No. 1509, p. 10, Springer–Verlag.

Ford, J. (1983), How random is a coin toss? *Phys. Today*, April 1983, p. 40.

Fredkin, E. and Toffoli, T. (1982), Conservative logic, *Int. J. Theor. Phys.* **21**, 219.

Furusawa, A., Sørensen, J. L., Braunstein, S. L., Fuchs, C. A., Kimble, H. J., and Polzik, E. S. (1998), Unconditional quantum teleportation, *Science* **282**, 706.

Galindo, A. and Martin-Delgado, M. A. (2002), Information and computation: Classical and quantum aspects, *Rev. Mod. Phys.* **74**, 347.

Garey, M. R. and Johnson, D. S. (1979), Computers and intractability: A guide to the theory of NP-completeness, W. H. Freeman, New York.

Georgeot, B. and Shepelyansky, D. L. (2000), Quantum chaos border for quantum computing, *Phys. Rev. E* **62**, 3504; Emergence of quantum chaos in the quantum computer core and how to manage it, *Phys. Rev. E* **62**, 6366.

Georgeot, B. and Shepelyansky, D. L. (2001a), Exponential gain in quantum computing of quantum chaos and localization, *Phys. Rev. Lett.* **86**, 2890.

Georgeot, B. and Shepelyansky, D. L. (2001b), Stable quantum computation of unstable classical chaos, *Phys. Rev. Lett.* **85**, 5393; see also *Phys. Rev. Lett.* **88**, 219802 (2002).

Gisin, N. and Massar, S. (1997), Optimal quantum cloning machines, *Phys. Rev. Lett.* **79**, 2153.

Gisin, N., Ribordy, G., Tittel, W., and Zbinden, H. (2002), Quantum cryptography, *Rev. Mod. Phys.* **74**, 145.

Gorbachev, V. N. and Trubilko, A. I. (2000), Quantum teleportation of EPR pair by three-particle entanglement, *J. Exp. Theor. Phys.* **91**, 894.

Gossett, P. (1998), Quantum carry–save arithmetic, quant-ph/9808061.

Gottesman, D. and Chuang, I. L. (1999), Demonstrating the viability of universal quantum computation using teleportation and single-qubit operations, *Nature* **402**, 390.

Grassi, A. M. and Strini, G. (1999), Some extensions of the Deutsch problem, in Mysteries, puzzles, and paradoxes in quantum mechanics, Bonifacio, R. (Ed.), AIP Conf. Proc. 461, p. 291.

Grover, L. K. (1996), A fast quantum mechanical algorithm for database search, quant-ph/9605043, in Proc. of the 28th Annual ACM Symposium on the Theory of Computing, p. 212, ACM Press, New York.

Grover, L. K., (1997), Quantum Mechanics helps in searching for a needle in a haystack, *Phys. Rev. Lett.* **79**, 325.

Gruska, J. (1999), Quantum computing, McGraw–Hill, London.

Gulde, S., Riebe, M., Lancaster, G. P. T., Becher, C., Eschner, J., Häffner, H., Schmidt-Kaler, F., Chuang, I. L., and Blatt, R. (2003), Implementation of the Deutsch–Jozsa algorithm on an ion-trap quantum computer, *Nature* **421**, 48.

Haake, F. (2000), Quantum signatures of chaos (2nd. Ed.), Springer–Verlag.

Hirvensalo, M. (2001), Quantum computing, Springer–Verlag.

Hughes, R. J. (1998), Cryptography, quantum computation and trapped ions, *Phil. Trans. R. Soc. Lond.* **A356**, 1853.

Jones, J. A., Mosca, M., and Hansen, R. H. (1998), Implementation of a quantum search algorithm on a quantum computer, *Nature* **393**, 344.

Jones, J. A. (2001), NMR quantum computation, quant-ph/0009002, in *Prog. NMR Spectr.* **38**, 325.

Jozsa, R. (1997), Quantum algorithms and the Fourier transform, quant-ph/9707033.

Jozsa, R. and Linden, N. (2002), On the role of entanglement in quantum computational speed-up, quant-ph/0201143.

Kane, B. (1998), A silicon-based nuclear spin quantum computer, *Nature* **393**, 133.

Kitaev, A. Yu. (1995), Quantum measurements and the Abelian stabilizer problem, quant-ph/9511026.

Klappenecker, A. and Rötteler, M. M. (2001), On the irresistible efficiency of signal processing methods in quantum computing, quant-ph/0111039.

Knill, E., Laflamme, R., and Milburn, G. J. (2001), A scheme for efficient quantum computation with linear optics, *Nature* **409**, 46.

Knuth, D. E. (1997–98), The art of computer programming, vol. I: Fundamental algorithms; vol. II: Seminumerical algorithms; vol. III: Sorting and searching, Addison–Wesley, Reading, Massachusetts.

Kurtsiefer, C., Zarda, P., Halder, M., Weinfurter, H., Gorman, P. M., Tapster, P. R., and Rarity, J. G. (2002), A step towards global key distribution, *Nature* **419**, 450.

Landauer, R. (1961), Irreversibility and heat generation in the computing process, *IBM J. Res. Dev.* **5**, 183.

Lang, S. (1996), Linear algebra, Springer–Verlag.

Lavor, C., Manssur, L. R. U., and Portugal, R., (2003), Shor's algorithm for factoring large integers, quant-ph/0303175.

Lee, J.-S., Chung, Y., Kim, J., and Lee, S. (1999), A practical method of constructing quantum combinatorial logic circuits, quant-ph/9911053.

Lloyd, S. (1995), Almost any quantum logic gate is universal, *Phys. Rev. Lett.* **75**, 346.

Lloyd, S. (1996), Universal quantum simulators, *Science* **273**, 1073.

Lo, H.-K., Popescu, S., and Spiller, T. (Eds.) (1998), Introduction to quantum computation and information, World Scientific, Singapore.

Lomonaco, S. J. (Ed.) (2000), Quantum computation: A grand mathematical challenge for the twenty-first century and the millennium, American Mathematical Society Short Course, American Mathematical Society.

Lomonaco, S. J. (2000), A lecture on Shor's quantum factoring algorithm, quant-ph/0010034.

Lomonaco, S., J. (2001), A talk on quantum cryptography, or how Alice outwits Eve, quant-ph/0102016.

Loss, D. and DiVincenzo, D. P. (1998), Quantum computation with quantum dots, *Phys. Rev. A* **57**, 120.

Mattle, K., Weinfurter, H., Kwiat, P. G., and Zeilinger, A. (1996), Dense coding in experimental quantum communication *Phys. Rev. Lett.* **76**, 4656.

Mermin, N. D. (2003), Lecture notes on quantum computation, available at http://people.ccmr.cornell.edu/~mermin/qcomp/CS483.html.

Mertens, S. (2000), cond-mat/0012185, in Computational complexity for physicists, *Computing in Science and Engineering* **4**, 31.

Merzbacher, E. (1997), Quantum mechanics, John Wiley & Sons, New York.

Miquel, C., Paz, J. P., and Perazzo, R. (1996), Factoring in a dissipative quantum computer, *Phys. Rev. A* **54**, 2605.

Miquel, C., Paz, J. P., Saraceno, M., Knill, E., Laflamme, R., and Negrevergne, C. (2002), Interpretation of tomography and spectroscopy as dual forms of quantum computation, *Nature* **418**, 59.

Miquel, C., Paz, J. P., and Saraceno, M. (2002), Quantum computers in phase space, *Phys. Rev. A* **65**, 062309.

Monroe, C., Meekhof, D. M., King. B. E., Itano, W. M., and Wineland, D. J. (1995), Demonstration of a fundamental quantum logic gate, *Phys. Rev. Lett.* **75**, 4714.

Nakamura, Y., Pashkin, Yu. A., and Tsai, J. S. (1999), Coherent control of macroscopic quantum states in a single-Cooper-pair box, *Nature* **398**, 786.

Nielsen, M. A., Knill, E., and Laflamme, R. (1998), Complete quantum teleportation using nuclear magnetic resonance, *Nature* **396**, 52.

Nielsen, M. A. and Chuang, I. L. (2000), Quantum computation and quantum information, Cambridge University Press, Cambridge.

Ortiz, G., Gubernatis, J. E., Knill, E., and Laflamme, R. (2001), Quantum algorithms for fermionic simulations *Phys. Rev. A* **64**, 022319; erratum *ibid.* **65**, 029902 (2002).

Palma, G. M., Suominen, K.-A., and Ekert, A. K. (1996), Quantum computers and dissipation, *Proc. R. Soc. Lond. A* **452**, 567.

Papadimitriou, C. H. (1994), Computational complexity, Addison–Wesley, Reading, Massachusetts.

Peres, A. (1993), Quantum theory: concepts and methods, Kluwer Academic, Dordrecht.

Pittenger, A. O. (2000), An introduction to quantum computing algorithms, Birkhäuser, Boston.

Preskill, J. (1998), Lecture notes on quantum information and computation, available at http://theory.caltech.edu/people/preskill/.

Raimond, J. M., Brune, M., and Haroche, S. (2001), Colloquium: Manipulating quantum entanglement with atoms and photons in a cavity, *Rev. Mod. Phys.* **73**, 565.

Reck, M., Zeilinger, A., Bernstein, H. J., and Bertani, P. (1994), Experimental realization of any discrete unitary operator, *Phys. Rev. Lett.* **73**, 58.

Sackett, C. A., Kielpinski, D., King, B. E., Langer, C., Meyer, V., Myatt, C. Y., Rowe, M., Turchette, Q. A., Itano, W. M., Wineland, D. J., and Monroe, C. (2000), Experimental entanglement of four particles, *Nature* **404**, 256.

Sakurai, J. J. (1994), Modern quantum mechanics (revised ed.), Addison–Wesley, Reading, Massachusetts.

Saraceno, M. (1990), Classical structures in the quantized Baker transformation, *Ann. Phys.* (N.Y.) **199**, 37.

Schack, R. (1998), Using a quantum computer to investigate quantum chaos. *Phys. Rev. A* **57**, 1634.

Schrödinger, E. (1952), Are there quantum jumps? Part II, *Brit. J. Phil. Sci.*, **3**, 233.

Shor, P. W. (1994), Algorithms for quantum computation: discrete logarithm and factoring, in Proc. of the 35th. Annual Symposium on the Foundations of Computer Science, Goldwasser, S. (Ed.), p. 124, IEEE Computer Society Press, Los Alamitos, CA.

Shor, P. W. (1997), Polynomial-time algorithms for prime factorization and discrete logarithms on a quantum computer, *SIAM J. Sci. Statist. Comput.* **26**, 1484.

Song, G. and Klappenecker, A. (2002), Optimal realizations of controlled unitary gates, quant-ph/0207157.

Sørensen, A. and Mølmer, K. (1999), Spin–spin interaction and spin squeezing in an optical lattice, *Phys. Rev. Lett.* **83**, 2274.

Steane, A. (1998), Quantum computing, *Rep. Prog. Phys.* **61**, 117.

Strini, G. (2002), Error sensitivity of a quantum simulator I: A first example, *Fortschr. Phys.* **50**, 171.

Tapster, P. R., Rarity, J. G., and Owens, P. C. M. (1994), Violation of Bell's inequality over 4 km of optical fiber, *Phys. Rev. Lett.* **73**, 1923.

Terhal, B. and DiVincenzo, D. P. (2000), Problem of equilibration and the computation of correlation functions on a quantum computer, *Phys. Rev. A* **61**, 022301.

Tittel, W., Brendel, J., Zbinden, H., and Gisin, N. (1998), Violation of Bell inequalities by photons more than 10 km apart, *Phys. Rev. Lett.* **81**,

3563.

Tucci, R. R. (1999), A rudimentary quantum compiler (2nd. Ed.), quant-ph/9902062.

Turing, A. (1936), On computable numbers, with an application to the Entscheidungsproblem, *Proc. Lond. Math. Soc. (2)* **42**, 230; correction *ibid.*, **43**, 544 (1937).

Vandersypen, L. M. K., Steffen, M., Breyta, G., Yannoni, C. S., Sherwood, M. H., and Chuang, I. L. (2001), Experimental realization of Shor's quantum factoring algorithm using nuclear magnetic resonance, *Nature* **414**, 883.

Vazirani, U. (2002), Quantum computing, lecture notes available at http://www.cs.berkeley.edu/~vazirani/.

Vedral, V., Barenco, A., and Ekert, A. (1996), Quantum networks for elementary arithmetic operations, *Phys. Rev.* **A54**, 147.

Vion, D., Aassime, A., Cottet, A., Joyez, P., Pothier, H., Urbina, C., Esteve, D., and Devoret, M. H. (2002), Manipulating the quantum state of an electrical circuit, *Science* **296**, 886.

Weihs, G., Jennewein, T., Simon, C., Weinfurter, H., and Zeilinger, A. (1998), Violation of Bell's inequality under strict Einstein locality conditions, *Phys. Rev. Lett.* **81**, 5039.

Weinstein, Y. S., Pravia, M. A., Fortunato, E. M., Lloyd, S., and Cory, D. G. (2001), Implementation of the quantum Fourier transform, *Phys. Rev. Lett.* **86**, 1889.

Weinstein, Y. S., Lloyd, S., Emerson, J., and Cory, D.G. (2002), Experimental implementation of the quantum baker's map, *Phys. Rev. Lett.* **89**, 157902.

Welsh, D. (1997), Codes and Cryptography, Oxford University Press.

Wiesner, S. (1996), Simulation of many-body quantum systems by a quantum computer, quant-ph/9603028.

Williams, C. P. and Clearwater, S. H. (1997), Explorations in quantum computing, Springer Telos.

Wootters, W. K. and Zurek, W. H. (1982), A single quantum cannot be cloned, *Nature* **299**, 802.

Yamamoto, T., Pashkin, Yu. A., Astafiev, O., Nakamura, Y., and Tsai, J. S., Demonstration of conditional gate operation using superconducting charge qubits, *Nature* **425**, 941.

Zalka, C. (1998), Efficient simulation of quantum systems by quantum computers, *Fortschr. Phys.* **46**, 877.

Zalka, C. (1999), Grover's quantum searching algorithm is optimal, *Phys.*

Rev. A **60**, 2746.

Zurek, W. H. (1991), *Phys. Today*, October 1991, p. 36; see also quant-ph/0306072.

Zurek, W. H. (2003), Decoherence, einselection, and the quantum origins of the classical, *Rev. Mod. Phys.* **75**, 715.

Index

adjoint operator, 67
algorithmic complexity, 33
amplitudes, 53
AND gate, 18
anti-commutator, 73
Aspect's experiment, 96
authentication, 193

backward sign propagation, 117, 141
baker's map, 168
basis for a vector space, 60, 69
BB84 protocol, 199
Bell
  basis, 118
  inequalities, 92
  measurement, 210
  states, 90, 118, 197
billiard-ball computer, 45
binary
  addition table, 17
  arithmetics, 17
  functions, 132
bit, 16
Bloch
  sphere, 102
  vector, 102
Boolean functions, 23
BPP class, 28
BQP class, 29

Cæsar cypher, 189
CARRY circuit, 138

Cauchy–Schwarz inequality, 59
cavity quantum electrodynamics, 182
CCNOT gate, 43
change of basis, 69
chaotic systems, 31, 168
characteristic equation, 66
Chernoff bound, 30
CHSH inequality, 95
Church–Turing thesis, 13
circuit model of computation, 15, 105
classical cryptography, 189
CMINUS gate, 116
CNOT gate, 42, 113
coin-tossing sequences, 33
collapse of the wave function, 82
collective attack, 201
commutator, 71
complete
  orthonormal set, 68
  set of vectors, 60
completeness relation, 63
complexity classes, 27
composite systems, 88
computational
  basis, 100
  complexity, 24
conservative systems, 31, 80
constant of motion, 80
control qubit, 113
controlled
  gates, 112
  phase shift gate, 115

253